Controlling Carbon and Sulphur

During the financial year 1996/7 the Energy and Environmental Programme is supported by generous contributions of finance and technical advice from the following organizations:

Amerada Hess ● Ashland Oil ● British Gas
British Nuclear Fuels ● British Petroleum ● Brown & Root
Department of Trade and Industry, UK ● Eastern Electricity ● Enron
Elf ● ENI ● Enterprise Oil Esso/Exxon ● LASMO
Magnox Electric ● Mitsubishi Fuels ● National Grid ● Nuclear Electric
Powergen ● Ruhrgas ● St Clements Services Saudi Aramco ● Shell
Statoil ● TEPCO ● Texaco ● Veba Oel

Controlling Carbon and Sulphur

Joint Implementation
and Trading Initiatives

Academic edition

Proceedings of the 10th RIIA/IAEE/BIEE
International Energy Conference

Chatham House, London, 5–6 December 1996

edited by Dean Anderson
and Michael Grubb

THE ROYAL INSTITUTE OF
INTERNATIONAL AFFAIRS
Energy and Environmental Programme

First published in Great Britain in 1997 by
Royal Institute of International Affairs, 10 St James's Square, London SW1Y 4LE
(Charity Registration No. 208 223)

Distributed worldwide by
The Brookings Institution, 1775 Massachusetts Avenue NW,
Washington DC 20036-2188

A catalogue record for this book is available from the British Library.

Paperback: ISBN 1 86203 096 0

The Royal Institute of International Affairs is an independent body which promotes the rigorous study of international questions and does not express opinions of its own. The opinions expressed in this publication are the responsibility of the authors.

Printed and bound by The Chameleon Press Ltd.

Contents

Contributors to the conference

Dean Anderson, Associate Fellow, Energy and Environmental Programme, Royal Institute of International Affairs (RIIA), UK

Peter Bailey, Stockholm Environment Institute, University of York, UK

Richard Baron, Energy and Environment Division, International Energy Agency, France

Carlton Bartels, Director, Cantor Fitzgerald Environmental Brokerage Services, USA

Prof. Dr Bert Bolin, Chairman, Intergovernmental Panel on Climate Change (IPCC), Sweden

Dr Richard Bradley, Senior Advisor for Global Change, Office of Policy and International Affairs, Department of Energy, USA

Berndt Bull, Deputy Minister, Ministry of the Environment, Norway

Jan Corfee Morlot, Head, Climate Change Group, OECD Environment Directorate, France

Dr Robert K. Dixon, Director, US Initiative on Joint Implementation (USIJI), USA

Harald Dovland, Head of the delegation to the Framework Convention on Climate Change, Ministry of the Environment, Norway

Raúl A. Estrada-Oyuela, Chairman, Ad Hoc Group on the Berlin Mandate (AGBM), United Nations Framework Convention on Climate Change (UNFCCC) and Ambassador, Embassy of Argentina, China

William L. Fang, Deputy General Counsel, Edison Electric Institute, USA

Dirk Forrister, Assistant Secretary of Energy for Congressional Public and Intergovernmental Affairs, Department of Energy, USA

Carlos Fortin, Deputy Secretary-General, United Nations Conference on Trade and Environment (UNCTAD), Switzerland

Dr Michael Grubb, Head, Energy and Environmental Programme, RIIA

Bill Hare, Climate Policy Director, Greenpeace International, The Netherlands

Prof. Dr Thomas Heller, Professor of Law, Stanford University, USA

Jørgen Henningsen, Director, Environmental Quality and Natural Resources, DG-XI, European Commission, Brussels

Dr Tim Jackson, Senior Research Fellow, Centre for Environmental Strategy, University of Surrey, UK

Michael Jefferson, Deputy Secretary General, World Energy Council, UK

Michael Jenkins, OBE, Chairman, Futures and Options Association, UK

Frank T. Joshua, Head, Greenhouse Gas Emissions Trading Project, UNCTAD, Switzerland

Dr Klaus Kabelitz, Head, Economic and Energy Industry Department, Ruhrgas AG, Germany

Brian J. McLean, Director, Acid Rain Division, Environmental Protection Agency (EPA), USA

Dr Bert Metz, Head, Delegation to the UNFCCC, The Netherlands

Sascha Müller-Kraenner, Head, International Programme, Deutscher Naturschutzring (DNR), Germany

John Palmisano, Director, Regulatory Affairs, Enron Europe Limited, UK

David Porter, Head, Association of Electricity Producers, UK

Dr Katsuo Seiki, Executive Director, Global Industrial and Social Progress Research Institute (GISPRI), Japan

Dr Chandra Sekhar Sinha, Tata Energy Research Institute of New Delhi, India

Dr Katsunori Suzuki, Director, International Strategy on Climate Change, Global Environment Department, Environment Agency, Japan

Michael J. Walsh, Senior Vice President, Centre Financial Products Limited, USA

Prof. Dr Jan-Olaf Willums, Director, World Business Council for Sustainable Development, Switzerland; and Professor, Norwegian School of Management, Sandvika, Norway

Li Yun, Energy and Environment Division, Energy Research Institute, State Planning Commission, China; and Chinese Academy of Sciences

Preface

The conference on 'Controlling carbon and sulphur: international investment and trading initiatives' held on 5 and 6 December 1996 was the tenth in the annual series of conferences organized by the Royal Institute of International Affairs in association with the International Association for Energy Economics and the British Institute of Energy Economics. It represented something of a departure from previous conferences that had addressed relatively broader international energy topics. We decided to focus on a more specific topic that was arising on the international energy agenda: the possible use of trading instruments as a mechanism for controlling energy-related emissions of sulphur dioxide (SO_2) and carbon dioxide (CO_2).

In parallel with the accumulating experience of SO_2 control, the international debate on climate change has gathered momentum towards more substantive control of CO_2 emissions than is embodied in the skeletal Rio Framework Convention on Climate Change. At present, major questions remain unresolved not only about the level of commitment to be pursued, but also about the overall architecture for implementing any commitments on CO_2 emissions.

An important strand of the debate has been over 'Joint Implementation' of commitments involving government-supported investments by OECD companies in developing countries, and the less formal framework of 'activities implemented jointly' agreed at the Berlin Conference in 1995. A number of governments – notably the US and Japanese – are now promoting investment programmes under this umbrella both with central and east European countries and with developing countries. The World Business Council on Sustainable Development also launched a major private-sector initiative in support of such activities.

Activities implemented jointly, formal 'Joint Implementation' and the use of tradeable emission permits are all potentially instruments of international environmental policy that could affect the nature and economics of international energy investments, especially when applied to carbon. All this gives rise to a number of questions that are of huge importance for environmental policy and the energy industries. What are the real lessons from the implementation of SO_2 controls to date, and how may these controls evolve in the future? How are the international negotiations on CO_2 proceeding and where are they likely to lead? What investment initiatives are countries and industries already embarking upon, and what lessons if any does the experience of SO_2 control provide for international CO_2 policies?

These questions underlay our decision on the topic of the conference. During the early stages of its organization, interest was boosted by the US statement in July 1996 that gave backing for the principle of legally binding emission targets, implemented using efficient and flexible measures 'such as joint implementation and emissions trading', with the United States indicating strongly that it believed that its experience with SO_2 trading was proving highly successful and its belief that CO_2 control could follow a similar model. The year also saw steady expansion of national and industrial joint implementation programmes, several of which were presented at the conference.

The conference turned out to be the best attended for many years, almost filling the newly expanded conference venue at Chatham House, with about 250 delegates from industry, government, environmental groups and academics. The extensive and high-level participation was testament to the growing interest in the topics covered, and we were honoured by consistently high-quality presentations. As the first major conference on this topic in Europe it brought the issues to a wide new audience. Generous sponsorship from UNCTAD, MITI, the Japanese Environment Agency, and the US Environmental Protection Agency and Department of Energy, enabled us to offer prices below normal commercial rates, and supported many places for participants from academia, developing countries, and environmental non-governmental organizations.

Full conference proceedings, containing all the material presented, were produced by the RIIA's conference unit and marketed at commercial rates. Because of the topicality and breadth of interest in the issues discussed, especially in the context of the ongoing negotiations on climate change, the Energy and Environmental Programme has produced this academic edition for a wider audience. It contains a selection of the papers, focusing on the specific experience with joint implementation and emissions trading, and initiatives relating to the wider application of such instruments, but omitting some of the broader discussion of climate change science and processes, commentaries and position statements, and papers for which only overheads were available. We believe it brings together a wealth of analysis and discussion of the specific experience and issues related to joint implementation and emissions trading for controlling SO_2 and CO_2 that has not hitherto been available in one document.

July 1997

Michael Grubb
Head, Energy and Environmental Programme
Royal Institute of International Affairs

Acknowledgments

We are very grateful to all those who presented papers and participated at this conference for the consistently high quality of presentations and discussions. We would also like to thank the RIIA's Conference Unit, especially Philippa Challen and Diana Bailey, for undertaking the administration and logistics of the event; and RIIA's Publications Department and Christiaan Vrolijk, our Programme's Research Assistant, for helping with the production of these proceedings.

We are also most grateful to the following for providing the financial support which made possible extensive participation from non-OECD countries and from outside the corporate sector:

UN Conference on Trade and Environment
US Environmental Protection Agency
US Department of Energy
Ministry of International Trade and Industry, Japan
Japanese Environment Agency

We are also grateful for the intellectual input of individuals in these and other departments. The content of these proceedings, of course, remain the responsibility of the editors and of the named authors.

July 1997   Michael Grubb

Abbreviations

AGBM	Ad Hoc Group on the Berlin Mandate
AIJ	Activities Implemented Jointly
ARAC	Acid Rain Advisory Committee
ATS	Advanced tracking system
CDQ	Coke dry quenching
CEMS	Continuous emissions monitoring systems
CEUM	Coal and Electric Utility Model
CFCs	Chlorofluorocarbons
CLRTAP	Convention on Long-Range Transboundary Air Pollution
COP	Conference of the Parties (to the FCCC)
DNR	Deutscher Naturschutzring
EITs	Economies in transition
EMEP	European Monitoring and Evaluation Programme
EPA	Environmental Protection Agency (US)
EPRI	Electric Power Research Institute
ERC	Emissions reduction credit
ESTs	Environmentally sound technologies
ETS	Emissions tracking system
GEF	Global Environment Fund
GETS	Global Environmental Trading System
GHG	Greenhouse gas(es)
GISPRI	Global Industrial and Social Progress Research Institute
IBACC	International Business Action on Climate Change
IEA	International Energy Agency
IETO	International Emissions Trading Organization
IIASA	International Institute for Applied Systems Analysis
IMACC	Inter-Ministerial/Agency Coordination Committee
IPCC	Intergovernmental Panel on Climate Change
IUEPP	International Utility Efficiency Partnership Program
IWG	Interagency Working Group

JI	Joint Implementation
MtC	Million tonnes of carbon
Mtoe	Million tonnes of oil equivalent
NEDO	New Energy and Industrial Technology Development Organization (Japan)
ODA	Official Development Assistance
QELRO	Quantified emissions limitation and reduction objective
RAINS	Regional Acidification Information and Simulation Model
RECLAIM	Regional Clean Air Incentives Market
SBI	Subsidiary Body on Implementation
SBSTA	Subsidiary Board for Scientific and Technological Advice
SOI	Statement of Intent
SSP	Second Sulphur Protocol
UNCTAD	United Nations Conference on Trade and Development
UNDP	United Nations Development Programme
UNECE	United Nations Economic Commission for Europe
UNEP	United Nations Environment Programme
(UN)FCCC	United Nations Framework Convention on Climate Change
USIJI	US Initiative on Joint Implementation
VOC	Volatile organic compound
WBCSD	World Business Council for Sustainable Development

Chapter 1

Introduction and conference overview

Dean Anderson and Michael Grubb

Introduction

The aim of the conference on 'Controlling carbon and sulphur: international investment and trading initiatives' was to explore recent experience with the use of emissions trading and international investment instruments for emissions control, and to set the emerging debate about the possible use of emissions trading for controlling greenhouse gases in the context of this experience and wider political developments. This overview summarizes the main points of the presentations and conference debates within such an overall context.

The theme is not one that could possibly have formed the topic for such a major international conference even a year or two ago, for the debate draws heavily on experiences of the past few years, and recent developments in the science and politics of climate change. In 1994, the UN Economic Commission on Europe's (UNECE's) Second Sulphur Protocol was signed in Oslo, introducing long-term and stricter limits on SO_2 emissions, with some scope for international flexibility in its implementation. In 1995, the first SO_2 ceiling established under the US Clean Air Amendment Act came into force, providing the first concrete test of its unique emissions trading system; in the same year in Berlin, the first Conference of Parties to the UN Framework Convention on Climate Change (UNFCCC) launched a new round of negotiations on specific commitments, together with an agreement on a pilot phase of 'Activities Implemented Jointly'.

In July 1996, spurred by the findings of the Second Assessment of the Intergovernmental Panel on Climate Change, and flushed with the perceived success of its SO_2 control system and other emissions trading programmes, the United States announced in Geneva that it now favoured legally binding controls on greenhouse gas emissions, 'implemented using flexible and cost-effective instruments such as emissions trading and joint implementation'. Six weeks after our conference was held, the United States submitted specific proposals on emissions trading, and on joint implementation, to be included in the negotiating text for the agreement that is due to be reached in Kyoto in December 1997 – proposals that rapidly gained support from Russia, New Zealand, Norway and Australia, among others. With the issue now firmly on the agenda for Kyoto, we have produced these proceedings, and this overview of the papers and discussions, for the benefit of all those interested in the ongoing debate.

In theory, 'market instruments' such as emissions trading and joint implementation have many attractions as ways of controlling harmful emissions. By concentrating on quantified limitations, in principle they ensure that specified environmental goals are met. By avoiding governmental mandate of the means by which such goals are achieved, they give maximum scope for the development and application of locally appropriate solutions. By allowing market-based flexibility, they offer an economically efficient solution, enabling goals to be met at least economic cost – or tougher goals to be set within the same cost. By unleashing market incentives, they give maximum encouragement for innovations that can further lower the cost of achieving environmental goals.

That at least is the theory. The practical experience with market-based instruments, which dates back a couple of decades, is more complex. In the 1990s, however, there have been growing success stories and gathering support for specific instruments, such as the emissions allowance trading system now implemented for sulphur control in the US. This conference was held not to debate the theoretical attractions, but to explore the practical experience, and the nature and context of new proposals – and the implications all this may have for the future use of such instruments to tackle issues such as CO_2 emissions.

One point to note is the profusion, and sometimes confusion, of terms relating to joint implementation and to emissions trading. Box 1.1 summarizes the main terms, noting in particular the cases where alternate uses can sometimes create confusion. Most notably, while the term 'emissions trading' is quite clearly understood as referring to a range of different possible institutional systems that allow quantified emission limitations to be

---

Box 1.1: Relevant terms under the Climate Change Convention

AIJ (Activities Implemented Jointly)	Greenhouse gas reduction projects implemented in some FCCC Parties (developing countries or transition economies) but funded by others (OECD countries), but without any 'crediting' of emissions. This is seen as a pilot or test phase of future 'joint implementation' (see Box 1.2).
Annex I	The industrialized countries undertaking specific commitments in the UNFCCC and expected to sign up to legally binding emission commitments under the Kyoto agreement.
QELRO	'Quantified emissions limitation and reduction objective' – FCCC language for greenhouse gas targets and timetables.
(UN)FCCC	(United Nations) Framework Convention on Climate Change

---

exchanged between different entities, the term 'joint implementation' (JI) has tended to assume different meanings in the US from previous usage in the EU. In the United States, JI refers strictly to a situation in which a sponsor country's government or firm subject to an emissions control may undertake emission-reducing investments elsewhere, with a host partner not subject to quantified emissions control; the investor may then claim some or all of the resulting emissions reduction as a *credit* towards its own obligation. In the EU, the term has hitherto been used more liberally, as a concept acknowledging the validity of joint implementation of commitments. Thus the Oslo Protocol enables two parties to share goals for achieving SO_2 reductions – a sort of informal emissions trading – and labels this joint implementation. In this report, we use principally the US definition, which is becoming the dominant usage in the context of the climate change negotiations.

The term 'activities implemented jointly' (AIJ) was given the same meaning by all speakers. The term was introduced at the first Conference of Parties to the Climate Change Convention in Berlin, where debates over JI (which was strongly opposed by most developing countries) resulted in a compromise decision to allow a pilot phase of activities in which sponsoring industrialized countries and companies would invest in mitigation

Box 1.2: Definition of terms

Allowance/Entitlement	(1) The total allowed emissions from a controlled entity, i.e. synonym of *quota* or *cap* (see below); (2) synonym of *permit* (see below) in the sense of a right granted to emit a defined quantity of the controlled substance during a defined period.
Banking	A system whereby emission *permits* not used by a polluter in any given compliance period can be saved for use in a future period.
Baseline	(1) In reference to modelling studies and *JI*: projection of emissions that would occur in the absence of abatement project or policy intervention; (2) in reference to *emissions trading*: criteria by which initial emission allowances are determined and/or allocated among individual sources.
Borrowing	A system whereby emission *quotas* or *permits* required by an emitter in any given compliance period can be borrowed from the quotas or permits available to it in a future period; the borrowing may carry an 'interest rate' or a 'penalty' (in the form of a tougher standard or constraint).
Cap	An aggregate emissions limit applying to a set of polluters – e.g., a legally binding emissions total for a given year or budget period for a nation, an industry, or other group.
Credit	Additional emissions *allowance* earned by a controlled entity when it invests in reducing emissions from an uncontrolled source. In the context of a trading system, this means a participating entity earning a *permit* or *quota* through an investment that reduces emissions from a source that is not part of the full trading system.
Emissions budget	Essentially the same as *quota* or *cap*, but employing a period of several years as the control period – e.g., 500 million tonnes of carbon (MtC) over five years.
Emissions trading	The buying and selling of emissions *quotas, permits* or *credits*, either directly between controlled emitters, or indirectly via intermediaries (brokers, exchanges, etc.).

Joint Implementation	(1) Common terminology under UNFCCC: a system whereby greenhouse gas reduction projects are implemented in one FCCC Party and funded from another; the sponsoring Party gains a *credit* allowing it to increase its own emissions. (2) European terminology referring to some other Conventions: informal system allowing countries to achieve a given total emissions objective jointly, with flexibility in the relative contribution of different Parties.
Offset	A quantified reduction in emissions achieved by an investment in an uncontrolled source which is credited to the investing entity (sponsor) against its reduction obligation (see *credit*).
Permit	A marketable instrument conferring the right to emit a quantified amount of a pollutant, e.g., one tonne of carbon per year. 'Permit' is the term which has tended to be used when industries or individual firms are the trading bodies.
Quota	The emissions total applying to each polluter – e.g., 100 MtC per year. In a tradeable system, if the polluter emits 110 MtC, it must purchase quotas equal to 10 MtC from other polluters; if it emits 90 MtC, it has quotas equal to 10 MtC available to sell. 'Quota' is the term which has tended to be used when national governments are the trading bodies.
COP	Conference of Parties (the supreme body of the Convention, which meets each year).
AGBM	Ad Hoc Group on the Berlin Mandate (the principal negotiating body).

projects in other countries (principally east European and developing countries), subject to host country approval and certain criteria. The term avoided the connotation of joint implementation of emission commitments (since the developing countries had no specific commitments), and explicitly stated that emission reductions would not be credited to sponsor governments. The experience will be reviewed by the year 2000, with most industrialized countries' governments clearly hoping that this will pave the way for an agreement on full 'Joint Implementation'.

Development and experience of sulphur legislation
SO₂ agreements in Europe

Harald Dovland, Head of the Norwegian delegation to the UNFCCC, reviewed the history of European emission reduction protocols covering NO_x, VOCs, and SO_2. The Helsinki Protocol of 1985, Sofia Protocol of 1988, and VOC protocol of 1991 all called for flat-rate reductions. In contrast, the Oslo SO_2 Protocol of 1994 was based on differentiated commitments and used a 'critical load' approach, meaning that countries initially were required to close by 60 per cent the gap in their respective levels of compliance. Differentiation was intended to achieve the lowest cost for Europe as a whole. Individual countries, which under the Treaty undertake varying levels of reductions, were willing to comply with a polluter-pays approach because they perceived it as more equitable than a flat-rate approach and because joint implementation (EU definition) was allowed.

Emission sources and pollution impact areas were 'mapped' by computer models and served as the basis for reduction allocations. Inputs into the models included inventories of current and projected emissions, cost curves, and long-range transport simulations. Computer models ('maps') were reluctantly accepted by the participants as the basis for reduction obligations in the absence of a better alternative.

So what lessons could be learned from this European experience that might help in designing a CO_2 trading system? Not many, actually. First, the Oslo Protocol system was not a market-based approach, though it did allow for 'JI' (EU definition). Second, the distribution of SO_2 sources and impacted zones across Europe was highly variable, requiring a mechanism addressing 'hot spots'. A true trading system requires that the impacts are not closely related to the location of emissions, something which occurs with CO_2, since concentrations are fairly evenly distributed globally.

Dr Tim Jackson, Senior Research Fellow of the Centre for Environmental Strategy at the University of Surrey, discussed in theoretical terms the economics of the Oslo SO_2 protocol. He asked whether there was enough economic incentive for potential sponsors to make substantial investments. Could, for example, Sweden benefit significantly from investments in SO_2 abatement in Estonia? A complicating factor not applicable to CO_2 emissions was that sulphur emissions cannot be traded on a one-for-one basis in Europe because of the spatial complexity of sulphur deposition. A formula could, however, be used to capture spatial differences between sponsor and host countries to estimate the economic feasibility of trading. The formula was used to calculate an 'exchange rate' representing the difference between the sponsor country's reduction target and the decrease in emissions achieved in the host country. Two conclusions could be drawn from this exercise. First, to generate sufficient incentive for substantial investment in JI, the difference in the marginal cost of abatement in the respective countries had to be greater than the difference in their exchange rates. Second, as investment in the host country abatement expanded, reducing the differential, the gains from JI diminished rapidly. The actual economics of such investments would depend on rules under negotiation and yet to be adopted.

The US experience with SO_2 emissions trading

The 1990 amendments to the Clean Air Act established the first national emissions trading programme in the United States. *Brian McLean*, Director of the Acid Rain Division of the Environmental Protection Agency (EPA), described the programme as the first national market-based approach to environmental management. In 1990 both carbon/energy taxes and command-and-control measures were considered politically unfeasible, the former because of voter resistance and the latter because of estimated high costs, prompting strong industry opposition.

Both *Dirk Forrister*, Assistant Secretary of Energy for Congressional, Public and Intergovernmental Affairs, US Department of Energy, and Mr McLean stated that while differences between sulphur and carbon emissions required somewhat different trading schemes, certain common elements did apply. Both sulphur and carbon trading required a binding target, and in both cases it was essential to design as simple a system as possible.

The main difference between carbon and sulphur was that sulphur could largely be controlled at the point of emission by a number of mature abatement technologies, while carbon was emitted by innumerable sources, large and small, requiring demand- as well as supply-side measures.

The goal of the initial (current) phase of the US sulphur emissions trading programme was to reduce emissions in the electric power sector by 8.5mt from their 1980 level of 17.5mt over the period 1995 to 1999. Only 263 boilers out of a potential 2,000 were initially made subject to emission reductions in Phase I, and an additional 182 units targeted for Phase II were participating voluntarily. Thus the current trading scheme involved only 445 emission sources, a relatively modest number in relation to the number of potential CO_2 sources within the same geographic area. In 1995, the 445 units reduced their SO_2 emissions to 5.3mt, well below the cap of 8.7mt; surplus allowances were 'banked' for future use. One of the characteristics of emissions trading schemes was that, assuming effective monitoring and enforcement, they achieved neither more nor less than the reductions they were designed to achieve.

To achieve the emission reduction goal, the designers of the programme first established an emissions cap for the controlled sources. A total of 8.95 million allowances, each of which authorized the emission of one ton of SO_2, were issued to participants each year. Plants were allowed to reduce their emissions either through improved efficiency, stack abatement, or lower utilization. The government allocated allowances up-front to sources based on historical utilization. New sources must acquire them from existing holders or from periodic auctions. Allowances could be traded to any party and could be banked for use in a subsequent year.

To date, 1,700 trades involving 30 million allowances had been made, including trades which utilities had made internally among their own plants. Before trading started, allowances were expected to sell for between $200 and $1,000 per ton. In the first year actual prices ranged from $250 to $300 and in 1996 had been below $100, reflecting lower than expected marginal abatement costs.

Brian McLean summarized lessons learned from implementing the US sulphur trading system which might streamline the design of carbon trading programmes:

- Simplicity of system design enabled a faster start-up and lower administrative costs.
- Allowance allocation was a highly political game.
- Once the system was in operation the government should not try to fine-tune the market or participate in trading.
- Allowance banking lowered compliance costs by affording companies a degree of flexibility as to when to make major capital investments.
- Trading actually did work to lower compliance costs. Annual aggregate abatement costs were about half as much as estimated ($2 billion versus $4 billion).
- The use of market instruments did in practice reveal true costs.
- An allowance system was not a replacement for standards which remained in place and serve as the foundation for trading. It became the means to achieve the standards. An essential element of the system was substantial penalties for non-compliance.
- Estimates were that the government's overall start-up and administration costs were less than half of what they would have been to manage a command-and-control system.

William Fang, Deputy General Counsel of the Edison Electric Institute, provided the perspective of the utility industry on the US experience with sulphur trading. He agreed fundamentally with the views of Brian McLean and Dirk Forrister regarding the benefits of trading in bringing about larger than expected emissions reductions at lower than expected costs. However, he argued that trading was 'not the biggest factor' in these achievements. Utilities would have pursued efficient alternatives under a traditional command-and-control system as well. With respect to lessons learned from the US sulphur trading experience:

- Joint implementation was possible without emissions trading but emissions trading without JI 'makes no sense'. The point here is that (under the 'US definition' of JI), credits would be earned by Annex I countries or companies from those countries, motivating them to invest in emission reduction projects in host countries. In contrast, emissions trading by definition took place among power plants or other point sources compelled to reduce their emissions. As non-Annex I parties to the Convention, host countries had no such obligations, nor did they impose them on their domestic companies, providing no basis for emissions trading.

• On the basis of the US experience with sulphur trading, one should expect an allowance system to require five or six years to set up. A credit system would avoid the 'feeding frenzy' inherent in allocating allowances under a fixed total.

Carlton Bartels, Director, Cantor Fitzgerald Environmental Brokerage Services, described his firm's four-and-a-half-year experience in trading emission allowances (prior to which it had been in the business of brokering debt instruments). With respect to the US sulphur trading programme, the electricity utilities 'embraced' the market following an initial period of 'denial'. Utilities had been accustomed to complying with regulations imposed by the government and had been allowed in most cases to pass the costs on to their customers. They now treated allowances as assets and made business decisions on how best to manage them. Trading in sulphur allowances was easy, involving only a phone call. Broker fees average one per cent of value and transaction costs were extremely low. In contrast, bilateral trading in emission reduction credits (ERCs), exemplified by the southern California programme, was more cumbersome, resembling a property transaction. Mr Bartels was optimistic that trading schemes could be devised for greenhouse gases and advised the designers to learn from experience and try to 'keep it simple'.

Michael Walsh of Centre Financial Products Ltd, a US firm which designs products for financial markets generally, recounted his experience in devising and trading instruments similar to ones which may be created for greenhouse gas (GHG) emissions trading. He found several features to recommend in the American SO_2 scheme: it guarantees compliance or over-compliance; the allocation method provoked no arguments over baselines; and transaction costs were low. The main lessons learned were: (1) do not insist on a perfect system, just a good one; (2) start with one gas and a limited number of countries, perhaps ten, and a finite time horizon, say ten years. One should be able to set the rules and design the system within two to three years.

The context for climate change

Scientific and energy dimensions

Bert Bolin, Chairman of the IPCC, described the essential differences between CO_2 and SO_2, saying that no critical load of CO_2 had been identified, and regional distribution of impacts was not well understood and thus could not serve as the basis for attribution to sources. The balance of Dr Bolin's talk was devoted to presenting some of the key findings of the 1995 IPCC Second Assessment Report with respect to the levels of reductions needed to achieve stabilization of greenhouse gas concentrations at levels which it might be hoped would not prove dangerous to human populations and ecosystems.[1] Rather than define 'safe' levels, which is ultimately a political determination to be made by delegates to the Framework Convention, Dr Bolin described emission levels needed to achieve stabilization of CO_2 concentrations at 450ppm and 550ppm respectively. This range made sense to participants in the negotiations because the IPCC's tentative findings on significant impacts related to concentration levels in excess of 550ppm, while 450ppm was often cited by environmental NGOs and delegates of small island states as an appropriate stabilization target limit, i.e. one offering hope of avoiding 'dangerous' impacts.

Since developing countries' emissions were increasing rapidly (by approximately 4 per cent per year) while developed countries' emissions were tending towards stabilization (increasing at the rate of 0.5 to 1.0 per cent per year), there was a clear need to bring about a transition in developing countries to cleaner technologies. Overall, the gap between the emission projections and the levels needed to stabilize the atmosphere was one that implied the need for binding targets, and flexible mitigation approaches.

Michael Jefferson, Deputy Secretary General of the World Energy Council, criticized portions of the IPCC's findings as premature, and concluded on balance that they may underestimate the scale of the problem.

---

[1] Dr Bolin's talk is not reproduced in these proceedings, which focus more specifically on the aspects of international negotiations and instruments. Discussions of the IPCC's Second Assessment Report are widely available, and include a Briefing Paper by RIIA's Energy and Environmental Programme, which presents a six-page summary of the Report and the debates underlying it (Duncan Brack and Michael Grubb, *Climate Change: A Summary of the IPCC's Second Assessment Report,* RIIA Briefing Paper No. 32, 1996).

As an indication of the direction and extent of emissions growth, he noted that between 1990 and 1995 the aggregate emissions of OECD countries increased by 4 per cent (with reductions in east Germany and the UK off-set by growth elsewhere) while those of developing countries increased by 25 per cent; emissions in central and east European countries declined. He criticized what he described as the unreality of much current discussion, which ignored the apparent inability of even most developed countries to get CO_2 emissions growth under control – most were failing to meet the target established by the Rio Convention, and some of those claiming to have done so had achieved it only through incidental gains, or through dubious statistical manoeuvres.[2]

He said that 'minimum regret' options were warranted, with immediate effect. Costs would be higher and dislocations greater and more unpredictable if action were put off. Specifically:

- Industrial countries should take the lead in transferring technology to the rest of the world.
- Capital stock turnover of emitting sources should be accelerated. However, in spite of efforts in this direction, it was important to realize that fossil fuels would continue to be used for power production for the foreseeable future.
- The rate of diffusion of renewable energy technologies should be increased.
- The rate of efficiency improvements should be increased.
- Expenditure on research and development should be increased.
- The public should be reassured about risks from nuclear energy.

---

[2] Most of the points made in Michael Jefferson's extensive paper are available in the series of Climate Change papers published by the World Energy Council, 34 St James's St, London WC1.

Li Yun, of the Energy Research Institute of the State Planning Commission of the People's Republic of China, gave an overview of efforts in China to reduce CO_2 and SO_2 emissions. Coal accounted for 75 per cent of primary energy supply today and this proportion was expected to decrease to 72 per cent, 68 per cent, and 63 per cent in 2000, 2010, and 2020 respectively. Sulphur emissions were expected to increase somewhat more slowly than carbon emissions, but China would still rank first in sulphur emissions by 2020.

Since the 1980s China had taken a number of steps to save energy and improve the environment. These included coal-washing, flue gas desulphurization, development of renewables (especially biomass and hydro), large-scale afforestation, and population control. In 1994 China adopted a 'Priority Programme' for meeting its medium- and long-term economic, social and environmental objectives, including its Agenda 21 commitments. These programmes contemplated various forms of international cooperation, including grants, loans, foreign investment, joint ventures, and build-operate-transfer (BOT) projects. Priorities for controlling sulphur and carbon included desulphurisation and dust removal, clean coal technologies, renewables, reclamation of wastes and mine tailings. The Ninth Five-Year Plan (1996–2000) included six demonstration projects in the areas of clean coal combustion, power plant efficiency, green lighting, and industrial heat recovery. In addition, the State Science and Technology Commission was developing solar, wind, biomass, geothermal, ocean thermal, hydrogen and fuel cell technologies. The Ministry of Coal Industry was developing low-sulphur coal mines and expanding coal-washing plants.

Finally, China was enlarging its use of foreign capital and encouraging joint-venture and cooperative projects to achieve its energy and environmental objectives.

The climate change negotiations

Ambassador Rául Estrada-Oyuela, Chairman of the Ad Hoc Group on the Berlin Mandate (AGBM), described the status and characteristics of the protocol negotiations. The negotiations were science-based, the parties to the Convention looking to the IPCC for guidance on the nature of the global warming phenomenon as well as on the risks it poses to human populations and eco-systems. In summary, he reported that:

- The Annex I parties (industrial countries) are generally failing to meet their CO_2 stabilization commitments.
- Any serious effort to force non-Annex I parties (developing countries), outside the OECD, to make emission reduction commitments would provoke an endless debate and could lead to destruction of the whole process.
- The Geneva Ministerial Declaration, endorsed by the vast majority of parties at the Second Conference of the Parties (COP-2) in July 1996, reaffirmed the commitment of the signatories to fulfilling the Berlin Mandate and to agreeing, by the end of 1997, an emissions reduction protocol for the post-2000 period.
- Some oil-exporting countries have disassociated themselves from the Declaration, making formal adoption of the document by consensus impossible. Thus the document was 'noted' rather than adopted by the Convention.
- National interests would continue to influence the debates.
- For the upcoming December negotiating session (AGBM-5) Chairman Estrada and his staff had prepared a synthesis of proposals from the parties on features of a protocol or other legal instrument. Every attempt was made to draft language reflecting accurately the proposals submitted.
- Proposals for quantified emissions limitation and reduction objectives (QELROs) varied significantly. Some mentioned specific targets and timetables. Some offered formulae for differentiating commitments. Some argued for uniform targets but differed as to whether they should apply to individual countries or to Annex I as a whole. Some suggested adopting targets initially for CO_2 only, while others suggested that targets should cover a 'basket' of gases.
- Proposals for policies and measures varied widely as well. The US delegation emphasized the need for flexibility to choose among proposals, while the EU recommended three groups of measures: (1) ones which would be 'common to' parties, i.e. mandatory; (2) ones which would be voluntary but require coordination amongst countries to be effective; and (3) ones which would be optional but recommended for domestic implementation.
- Probably not all countries would join the protocol. This raised the ques-

tion of what to do about free-riders, which would need to be addressed since they could not be given a 'blank cheque' to frustrate the purposes of the Convention.

- Many no-cost and low-cost measures were available to help achieve emission reductions, but not all costs could be avoided. The Global Environment Fund (GEF) will continue as the main mechanism for assistance to developing countries in meeting their greenhouse gas inventory and reporting commitments.

Political dimensions

Dirk Forrister described the Clinton administration's strong support for greenhouse gas emissions trading in the context of both the 'evolution' of climate change policy in the administration's first term and the positive experience of the government with emissions trading, especially sulphur. The policy evolution took place during two years of contentious politics between the White House and Capitol Hill. It emerged from the polls during the 1995–96 budget battle that a majority of voters believed the Republican Congress had gone too far in dismantling existing environmental laws. This prompted the Clinton/Gore campaign to treat the environment as one of its four main themes.

However, the election left a small Republican majority in the Senate, and the administration was concerned about its ability to get a CO_2 emissions reduction protocol ratified. Assuming all 45 Democrats voted in favour, it would need 22 Republican votes, which meant it must find a bipartisan basis for moving forward. The administration saw emissions trading as an essential element of its programme because trading was perceived as being more efficient than regulation and more politically realistic than carbon or energy taxes. Moreover, the positive experience of the government with sulphur trading provided a precedent for carbon trading.

Berndt Bull, Deputy Minister, Norwegian Ministry of the Environment, explained Norway's interest in both JI and emissions trading. This interest derived from Norway's national circumstances as a country with relatively low emissions (owing to its exclusive use of hydro for electricity production) and correspondingly high marginal cost of CO_2 abatement. Norway was a proponent of differentiated QELROs, both on equity grounds and because it believed differentiated commitments combined with flexible implementation mechanisms would in the long term be more globally efficient than a flat-rate approach. Mr Bull's more detailed comments on Norwegian JI programmes and on emissions trading are summarized below.

Carlos Fortin, Deputy Secretary-General of UNCTAD, provided a complementary political perspective. Sustainable development was the goal of developing and developed countries alike. Both were interested in market-based approaches, the former because of their limited financial resources and administrative capacities, and the latter because of the proven cost-effectiveness and likely trade benefits of emissions trading if implemented globally.

UNCTAD recognized the need for developing countries to develop their economies in order to solve their pressing environmental and social problems and to improve the living standards of their populations. However, in order to be effective, any international climate change agreement must include abatement in developing countries, where emissions growth was rapid. Therefore what was needed was a mechanism for allowing development to take place using efficient, environmentally friendly technology.

In 1990 UNCTAD became interested in the potential for emissions trading to help achieve these broader developments as well as environmental goals and initiated the first of five studies, leading it to advocate a pilot programme involving developing countries and economies in transition (EITs) as well as developed countries. While deferring comment on technical aspects of trading to Frank Joshua's subsequent presentation, Mr Fortin offered some general conclusions:

- The US experience showed trading to be efficient.
- Trading was neither a licence to pollute nor inherently unfair or unworkable.
- Trading could potentially achieve annual North to South investments of $40–50 billion annually, the equivalent of all development aid today.

Joint implementation and activities implemented jointly
National JI programmes of OECD countries

Dr Robert Dixon, Director, US Initiative on Joint Implementation (USIJI), described this programme, which was launched in 1993 as part of the US Climate Action Plan. The objectives of this AIJ programme were to:

- promote a broad range of projects in order to evaluate a variety of carbon reduction measures;
- help host countries to meet the objectives of their climate action plans and to develop their monitoring and verification capabilities;
- encourage private-sector investment;
- achieve efficient reductions;
- complement the US country studies programme;
- 'enrich' the AIJ pilot phase approved at COP-1 by providing concrete evidence of the contribution which cooperative mitigation projects can make to global emissions reduction.

The USIJI programme was expanding; a new 'batch' of projects had just been approved, increasing the total number of projects 'accepted' during the first three years of the programme to 24. In spite of the unavailability of crediting in the pilot phase, interest from the business community had been increasing, as evidenced by the fact that 75 proposals meeting the government's basic criteria were received in the latest (third) round (eight of these were accepted). The basic criteria are:

- host country acceptance;
- specific measures to reduce or sequester emissions;
- additionality (emissions reduction and funding must be additional to investments which would have been made for other investment-related reasons);
- identification of non-GHG environmental and socio-economic impacts;
- provision for adequate monitoring, verification, and reporting.

Dr Dixon referred to two typical projects:

- a district heating system in the Czech Republic involving fuel-switching from coal- to gas-fired heat generation;
- development of a new power plant in Honduras which would be fired by wood waste, thus displacing some diesel fuel use for electricity generation.

Finally, he mentioned that sponsors of the latest round of projects had had an easier time obtaining financing than their predecessors, even though the projects were larger than those accepted in the previous two rounds. This implied increased interest in AIJ projects on the part of investors and lenders.

Bert Metz, Head of the Dutch delegation to the UNFCCC, gave an overview of the 20 or so AIJ projects in central and eastern Europe. Some west European countries opposed the concept of JI and were not sponsoring any projects; some projects involved technology transfer and were subsidized by governments; most projects were energy-related, though there were also a few sequestration (forestry) projects. The project sector in general had been lukewarm to the concept because of the lack of crediting.

Dr Metz mentioned a few examples of projects: Norway was sponsoring a coal to gas fuel-switching project in Poland; Sweden was developing a wind farm in Latvia; Finland had an energy efficiency project in Russia; and Denmark had gas-fired district heating projects under way in Poland and the Czech Republic. (Dr Metz did not specify whether these projects were initiated and financed by the private or the public sector.) The government of the Netherlands had established a $50 million fund through 1999 to assist private sector-sponsored AIJ projects. The Electricity Board had also established a fund. The government required AIJ projects to be economically sound and to offer financial 'additionality'. This definition of additionality, one of several uses of the term, meant that funds additional to existing government aid and private investment flows were earmarked specially for purposes of emissions reduction.

In the protocol negotiations the Netherlands supported the availability of JI sponsorship opportunities, with crediting, to countries agreeing to be bound by emission reduction obligations.

Dr Katsunori Suzuki, Director of the Office of International Strategy on Climate Change of the Japanese Environment Agency, provided a brief overview of the 'Japanese Programme for AIJ under the Pilot Phase'. The objectives of the programme were to gain experience, establish a methodology for measuring reductions, and encourage private-sector involvement. The government issued guidelines for project proposals in January 1996 and began receiving applications in April. In July it identified eleven potentially promising projects, which included one combustion improvement project, two power-sector energy-efficiency projects, one solar project, one coke dry quenching project, and six afforestation projects.

The government had established the Inter-Ministerial/Agency Coordination Committee for AIJ (IMACC) to set guidelines, evaluate project applications and performance, encourage private-sector participation, and report on the progress of the AIJ programme. However, it was too early to draw conclusions.

Dr Katsuo Seiki, Executive Director of the Global Industrial and Social Progress Research Institute (GISPRI), described three Japanese energy-related AIJ projects authorized by the Ministry of International Trade and Industry (MITI) and sponsored by the Japanese New Energy and Industrial Technology Development Organization (NEDO):

- A project in Thailand aimed at improving the thermal efficiency of a power plant. The principal participants from Japan were Kansai Electric Power Company, Chubu Electric Power Company, and Electric Power Development Company.
- Local electrification in Indonesia. This was a joint project involving the major power companies of the G-7 countries. The plan was to install home-use solar batteries (around 50kW), mini hydro plants (around 200kW), and hybrid solar battery with diesel backup units in cooperation with Indonesian utilities.
- A coke dry quenching project in China. NEDO was the main implementing entity. Dry quenching avoided the use of cooling water which caused pollution as steam was wasted into the atmosphere. Reclaiming the heat also saved energy.

Dr Seiki expressed the personal view that JI could make binding emission reduction targets for Annex I countries acceptable. A big problem at present was the lack of financial incentive, causing industry to be lukewarm about the concept. The incentives available today – public relations benefits and the opportunity to gain experience – were simply not enough. Dr Seiki supported the idea of an AIJ promotional forum in which representatives of national and local government agencies and industrial and environmental organizations could get together to exchange views.

Berndt Bull emphasized that Norway supported the concept of incorporating JI with crediting in an emissions reduction protocol for the post-2000 period and was eager to see measuring, monitoring and reporting criteria developed to make this possible. It advocates JI with crediting within Annex I. In cooperation with the World Bank, UNEP and UNDP, Norway is co-funding two AIJ pilot projects:

- in Poland, the conversion of several coal-fired power plants to gas; and
- in Mexico, promotion of efficient lighting in two cities.

New projects in Costa Rica and Burkina Faso were planned and efforts were being made to identify possible AIJ projects in eastern Europe and the Baltic states. Norway was working with the World Bank and the other Nordic countries to identify opportunities and facilitate learning by doing.

Implementation and politics of JI

Dr Thomas Heller, Professor of Law at Stanford University, discussed the politics of JI in the negotiations.[3] The early emphasis of developing countries was on the need for industrial countries to reduce their domestic emissions, since roughly three-quarters of the problem could be attributed to them. JI appeared at the time to be a diversion. More recently the focus had been shifting from historical responsibility towards pursuing cost-effective emission reduction paths. A problem was that no one had a good sense of what parties to the Convention will be willing to pay for mitigation. Dr Heller thought damage would be distributed differentially – countries which perceived the possibility of realistic adaptation options would be willing to pay less. Countries would be deterred from 'crossing over to an adaptation

---

[3] For a more detailed discussions see Thomas C. Heller, 'Environmental Realpolitik: Joint Implementation and Climate Change', *Indiana Journal of Global Legal Studies*, Vol. 3, Issue 2, spring 1996.

strategy' to the extent that the international community was able to find cheap reductions paths. An example was infrastructure development opportunities in China and other rapidly developing countries, opportunities unique to their present situation (early stage of development) that would not be available indefinitely. Investments in new technology now would avoid more expensive retrofits in the future. JI offered a unique chance to lead into full emissions trading, which is the direction Dr Heller advocated.

Prof. Dr Jan-Olaf Willums, Director of the World Business Council for Sustainable Development (WBCSD), depicted JI as having got off to a slow start because of the perception of many developing countries that it was a ploy by industrial countries to avoid expensive domestic emission reduction measures. A 'transfusion' was needed in the forms of: (1) a change in the conceptual framework from one of 'taking' to one of 'ownership'; and (2) a business-oriented approach offering sufficient incentives to mobilize private-sector resources. Annex II parties, as sovereign states, should view carbon offset opportunities as 'mineable resources' which could be left in the ground or exploited *at their discretion and under terms acceptable to them.* They could either grant development concessions or trade development rights for access to technology. This could lead to a simple carbon trading scheme crediting investors for emission reductions.

Dr Willums drew a distinction between technology transfer and technology cooperation. The former was often associated with government-to-government programmes, many of which have involved less-than-ideal technology in terms of its environmental characteristics and appropriateness for sustainable development. An example was 'moving' a steel plant from North to South or transferring existing steel production technology. Technology cooperation would involve joint efforts by sponsor and host country *companies* to adapt or develop new technology to fit local socio-economic and environmental circumstances. It was more important to transfer methods than hardware and essential to recognize and mobilize local skills. Dr Willums mentioned an AIJ project in Russia in which local expertise in boiler technology contributed significantly to a combined cycle gas turbine installation. It should also be kept in mind that governments have limited funds and do not control technology.

Subsidies were less important to investors than government assistance with risk management.

Dr Willums mentioned efforts by the WBCSD to encourage developing countries and their businesses to initiate partnership JI projects. It had recently launched the International Business Action on Climate Change (IBACC) initiative, the purpose of which is to generate project proposals from developing countries. To date 80 'proper' proposals had been received, of which 30, representing a potential total investment of $293 million, appeared to be 'realistic'. The World Bank was assisting with project evaluations. But for JI to make a significant impact on climate change, crediting would be required, as well as clear methods for measuring and monitoring emission reductions.

Dr Willums was asked whether he thought the Chinese government would be motivated to invite foreign investors to make investments in 'additional' GHG reduction. In response he gave the hypothetical example of a foreign automobile manufacturer faced with the option of building in China a more CO_2-friendly but also more expensive manufacturing plant. JI would offer a helpful additional incentive but not drive the decision.

Dr Chandra Sekhar Sinha, of the Tata Energy Research Institute of New Delhi, expressed some concerns which developing countries have about AIJ and JI:

- No ODA (overseas development assistance) funds should be diverted to finance AIJ projects;
- AIJ project financing should be supplementary to GEF funding;
- AIJ projects should only be allowed if they offer lower-cost emission reductions than *all* available host country opportunities;

JI should be implemented in three stages. During the first stage, corresponding to the present AIJ pilot phase, no crediting of emission reductions should be allowed either prospectively or retroactively. During the second stage (corresponding to the period of time during which developing countries have not yet adopted binding QELROs), crediting should be allowed; in the final stage (after the developing countries have agreed QELROs), they would enter the international emissions trading regime.

Emissions trading for greenhouse gases

Proposals for international emissions trading for greenhouse gases

Frank Joshua, Head of the Greenhouse Gas Emissions Trading Project of UNCTAD, summarized the early results of the five UNCTAD studies on emissions trading published between 1992 and 1995:[4]

- Trading could achieve emission reductions earlier and at lower cost.
- Trading was more politically acceptable than taxes and command-and-control alone.
- An allocation method could be selected to address equity issues. The urgent development needs of poorer countries could be reconciled with the global need to reduce greenhouse gas emissions.
- While a larger share of 'entitlements' could be given to developing countries to address social and economic problems, the introduction of per capita allocation criteria served only to polarize the discussion.
- Trading was 'win-win' for buyers and sellers. Buyers won by achieving a given reduction target at lower cost, sellers by being given maximum flexibility to choose how to accomplish abatement.
- The credibility of the trading programme rested on the integrity of monitoring, certification, and enforcement.

With respect to an eventual global CO_2 emissions trading system, UNCTAD believed that it should operate within the UNFCCC framework and that only countries adhering to QELROs should be eligible to participate. This meant Annex I parties plus any non-Annex I parties electing to participate. Crediting must be allowed if the system is to work. A gradual phase-in was needed to allow learning by doing. Trading should coexist with other mechanisms within countries. Only homogeneous gases should be traded, starting with CO_2, but trading in other greenhouse gases should be phased in.

Mr Joshua described the features of a possible phase one (2000–2004):

---

[4]   For the full set of references see Chapter 15.

- A new binding protocol under the UNFCCC would provide the basis for a voluntary 'group agreement' among countries adhering to QELROs. An international emissions trading organization would be created as a division of the UNFCCC. The IPCC could serve as the monitoring institute.
- The trading could be self-financing and set up within two years at a cost of under $10 million.
- Trading would take place within the private sector. Initially the trading system would be limited to major fixed energy sources.
- At the outset an expansion schedule would be agreed in order to minimize carbon leakage (circumvention).
- Initially trading would be in allowances rather than credits because the latter are more politically controversial since they avoid the setting of caps. (John Palmisano subsequently argued that, on the contrary, allowance trading was more political because it started with an allocation of permits, thereby creating assets over which participants quarrelled, whereas credits only came into being when a participant had earned them by reducing its emissions more than required; see below.)

Dr Richard Bradley, Senior Advisor for Global Change, Office of Policy and International Affairs, US Department of Energy, presented the general case for emissions trading versus taxes or regulation. First, trading offered greater certainty of achieving the desired environmental outcome than taxes. The result of taxation could be distorted by domestic politics and the granting of exemptions, subsidies, and the like. Second, trading reduced the cost of emissions reduction by allowing investment to flow to the least expensive options. Third, flexibility as to when and where to abate could save the world 'trillions'. The benefits of flexibility were recently shown in a study by the US electricity industry's Electric Power Research Institute (EPRI). The study assumed a 1990/2000 stabilization target and a 20 per cent reduction target for Annex I countries by 2000/2010, with trading taking place between high-cost and the other low-cost countries. Much of the trade was between OECD and former Eastern-bloc countries, including Russia. Four scenarios were analysed:

- in the first, regional targets were treated as fixed commitments, i.e. there was no flexibility to trade emission quotas, either internationally ('where' flexibility) or between successive periods ('when' flexibilty).
- the 'where flexibility' scenario saved 30 per cent worldwide relative to the no flexibility case.
- the 'when flexibility' scenario saved 70 per cent.
- the 'where/when flexibility' scenario saved 85 per cent, resulting in savings of between $2 and $8 trillion.

Dr Bradley believed trading should *facilitate* the negotiations by concentrating attention on opportunities for emissions reduction and technology diffusion and away from divisive issues such as historical responsibility for GHG concentrations and the gap in per capita emissions between industrial and developing countries. At present there was no effective means to bring about technology transfer. Annex I countries were being asked by non-Annex I countries to make expensive reductions at home and to transfer technology at below market rates. Political reality and governmental financial constraints made this infeasible, whereas emissions trading offered the opportunity to mobilize vast private-sector financial resources.

Dealing with different country circumstances would be challenging but manageable. For some countries trading among national governments might be the most workable approach. For others trading among and within companies might be more suitable. A mixed international system was needed, not only with respect to *who* traded but also with respect to *what* type of instruments were traded. Credit and allowance systems might coexist, at least in the beginning.

The US believed that in the post-2000 period, JI with crediting should be an option open to all countries. There was no reason for requiring countries to join Annex I in order to be eligible to participate. JI was also very important to the US as a means to achieve cost-effective reductions and to garner political support for undertaking binding QELROs.

The initial allocation of credits or allowances would be difficult but there were guiding precedents, including the Montreal Protocol, the US NO_x and SO_2 trading schemes, and UNFCCC stabilization targets (which allow some variations in setting base years). Domestic allocations might actually prove more difficult than international ones. Monitoring, verification and enforcement would be very important to maintaining integrity.

Jørgen Henningsen, Director of Environmental Quality and Natural Resources at DG-XI, European Commission, argued that the US sulphur trading scheme would not have been appropriate for Europe for two reasons: (1) because Europe has had better regulatory policies longer, which meant few 'cheap' opportunities remained which could drive active trading; and (2) trading only worked in areas where the geographical spread of emission effects was fairly even.

Comparing the preliminary lessons learned from US and European sulphur abatement experiences, one could conclude that emissions trading becomes interesting when expensive measures are required. So far no country needed to consider expensive measures relating to CO_2 emissions because of all the no- and low-regrets options available. For example, taxes could still reduce fuel consumption at no long-term cost to the vehicle owner (by inducing vehicle efficiency improvements and altering consumer choices). Many costs, such as subsidies, were *political*, induced by special interests. There is some scope for trading between OECD countries and former Eastern-bloc countries, but trading was not necessary to induce them. To the extent that in the US trading is used to avoid reductions, it would be a diversion from the purposes of the Convention.

Trading would become advantageous when we needed to come to grips with expensive reductions and once we were convinced of the commitments of the industrial countries to change their energy consumption patterns. Trading could only take place between countries with commitments; a scheme could theoretically be structured within Europe, but Mr Henningsen did not envision European governments or companies accepting allocations implying wealth or funds transfers from one country to another.

Berndt Bull described Norway's interest in emissions trading, which, it thought, had the potential to enhance the overall effectiveness of global efforts to reduce GHG emissions. Norway believed that the Kyoto agreement should stipulate flexible mechanisms for achieving QELROs, including emissions trading. The experience gained from the US sulphur trading scheme might hold valuable lessons on how to design a trading system; however, there was need for additional practical experience in more countries.

Norway was considering some form of trading as a means to assuring its compliance with the Oslo sulphur protocol. At present its two main instruments were a tax on the sulphur content of mineral oils and a licensing system for industry emissions. A decision would be made in the near future.

Dr Michael Grubb, Head of the Energy and Environmental Programme at the Royal Institute of International Affairs, discussed the merits of a 'cap and trade' system versus a tax system. Some economic considerations might favour a tax, while political factors point to 'caps'. While taxes were universally unpopular, cap allocations created an asset that could be reflected in corporate value, and could allow governments to buy off politically powerful industries through the number of permits initially allocated to them. Also issuing and administering permits (as opposed to taxes) fell more naturally into the purview of environmental ministries, not taxing bodies.

Dr Grubb suggested that an international emissions trading system for greenhouse gases could start with CO_2 and then gradually extend to other gases.

The effectiveness of emissions trading in reducing GHG emissions would vary considerably by country. This implied limits to the extent to which the US sulphur trading experience could be internationalized. One difference was the structure of emissions: none of Norway's and only 13 per cent of Switzerland's GHG emissions came from power plants or heavy industry. In these countries other forms of trading as well as other measures would be needed.

These factors raised the question of whether trans-European trading made sense. On balance, Dr Grubb believed it did, for several reasons:[5]

- the failures of EU-level policies, including the inability to impose a Community-wide carbon/energy tax and resistance by members to efficiency standards and other harmonized measures on subsidiarity grounds;
- liberalization of markets, imposing limits on what individual countries could do by way of imposing regulations and taxes;

---

[5] For a much fuller discussion see M. Grubb et al., *Implementing the European CO₂ Commitment: A Joint Policy Proposal,* London, RIIA, second edition, 1997.

- the exhaustion of low-cost measures;
- legal and institutional mechanisms in place, making a trading regime feasible if the political will could be mobilized; and
- international politics, which would require Europe to clarify exactly how it could deliver on its commitments.

The last point on this list was critical because the EU had presented itself as a leader in the protocol negotiations while at the same time appearing poised to fail to deliver on its 1990/2000 CO_2 stabilization commitment. Thus at the moment it was 'walking naked into the negotiating chamber'. A lead country was needed to spearhead interest in pan-European emissions trading. As suggested by Dr Bradley, a system of mixed (public-sector/private-sector) participation could be used, predicated on government quotas. It would be in everyone's best interest to include Norway and Switzerland. If the EU did follow this route, then emission allowance allocations would get bound up in EU structural and enlargement negotiations in the 1998–2000 time-frame.

With respect to the relationship of emissions trading and JI, it would be unwise to reopen the JI crediting issue before and up to the Kyoto conference. Wider joint implementation should be 'built on' successful demonstration of the concept in central and eastern Europe.

More broadly, the Kyoto agreement did not need to define or control national implementation systems. With respect to international trading, all that was needed was to define the national obligations and the circumstances under which trading was to be allowed. Through the rest of the AIJ pilot phase, JI should be defined as applying within Annex I (as it is now).

Issues in implementing greenhouse gas emissions trading
Richard Baron of the Energy and Environment Division of the International Energy Agency provided an overview of the conclusions of the Annex I Experts Group's study on global emissions trading.[6] He

---

[6] Fiona Mullins, OECD, and Richard Baron, IEA, *International GHG Emission Trading, Policies and Measures for Common Action*, Working Paper 9, Annex 1, Expert Group on the UNFCCC, OECD/IEA, Paris, March 1997.

described the various options available, which include trading within countries (by companies or emission sources), trading among governments or companies internationally, trading in credits or allowances, and trading within Annex I or between Annex I and non-Annex I countries. Though many forms were possible, governments should administer and monitor the programme since they make treaty commitments and had sovereignty. For a trading system to work there must be a constraint on emissions. Some form of allocation (of permits or credits) was implied. The system should be designed with flexibility to introduce new gases and bring in new parties over time. However, rules need to be certain and continuous to uphold the value of the instrument traded.

John Palmisano, Director of Regulatory Affairs, Enron Europe Limited, argued for emission reduction credit (ERC) rather than allowance trading. The US sulphur allowance trading programme took seven years to set up and was governed by 1,100 pages of regulations. Yet it applied initially to only a few hundred emission sources. Allowance allocations were highly political and invited 'grandfathering' claims from powerful interests. An international programme of significant size would be unmanageable both administratively and politically.

The (simplified) sequence in allowance trading was:

- establish baseline (allocate allowances);
- sell CO_2 emission reductions (trade first);
- document reduction;
- demonstrate compliance (comply second).

The steps in ERC trading were:
- establish baseline;
- document credit generating activity (comply first);
- get regulatory quantification and certification;
- either sell or bank ERCs (trade second).

Mr Palmisano envisioned trading by fuel companies (buyers), renewable energy project developers (sellers), intermediaries (brokers, speculators, hedgers), and some governments. Fossil-fuel companies, which had an

innate long position in carbon, would 'sell short'. Technology companies
would gamble on their ability to reduce and thus earn credits. If it was to
work, the system would need 'bullet-proof' integrity. It should first prove
itself in Annex I before expanding to the rest of the world. Eventually a
great deal of trading would take place *within* multinational companies.

Following Mr Palmisano's presentation, the question was raised
whether crediting would be allowed within Annex I prior to the completion and review of the AIJ pilot phase at the end of 1999. Bert Metz
answered that it *might* be allowed. Frank Joshua said that according to 'the
lawyers' it was not allowed; Chairman Estrada said there had been no specific discussion in the negotiations on this issue.

Jan Corfee Morlot, Head of the Climate Change Group in the OECD
Environment Directorate, gave her personal views on 'cap and allocate'
versus 'baseline and credit' systems. Phased-in CO_2 ceilings were necessary for either approach to work. A baseline and credit system might be
predicated on a 'with policies' baseline or a 'without policies' baseline;
however, the choice would affect the outcome and should be made with
this in mind. Credits were earned for reductions in excess of those required
to remain under the ceiling.

Reactions of non-governmental groups

Sascha Müller-Kraenner, Head of the International Programme of the
Deutscher Naturschutzring (DNR) explained that it was the role of environmental NGOs to focus on environmental objectives. DNR supported
the emission reduction target proposed by the Association of Small Island
States (AOSIS) – 20 per cent reduction in CO_2 by Annex I parties between
1990 and 2005. If the protocol negotiations produced a commitment on the
part of Annex I countries to achieve the target, DNR would be 'flexible' on
policies and measures. Emissions trading appeared to be an interesting
instrument to complement eco-taxes at the national level.

Ratification by the US Senate was being represented by some as a political reality 'test' which proposals for QELROs and policies and measures
must pass. However, public opinion was also a test, and many people
viewed polluting and buying one's way out as immoral. But public dis-

cussion of the climate change issue was not yet very advanced – the public was not well-informed or actively engaged, but this would change.

Good reasons for allowing emissions trading would be *if* it helped reduction obligations to be met efficiently and *if* developing countries could be convinced of its benefits.

DNR had developed a list of possible criteria for environmental NGO support for the concept:

- Only parties agreeing reduction obligations should be allowed to participate.
- QELROs should be the basis for the allocation of allowances or credits (certificates).
- As the parties to the protocol, national governments must be ultimately responsible for meeting QELROs, irrespective of who the parties were to emissions trading transactions.
- Rules must be adopted to prevent double-counting of credits and inflated baselines.
- Allocation of allowances or credits should be matched to QELROs timetables and reflect progressive strengthening of reduction targets in regular stages.
- QELROs must be backed by a credible international authority with real enforcement powers.
- Emissions trading should be part of a mix of policy instruments and not displace other measures, including taxes.

Bill Hare, Climate Policy Director of Greenpeace International, presented Greenpeace's views on emissions trading. Environmentalists were sceptical but open-minded about emissions trading, while both environmentalists and consumers were sceptical about the alleged high cost to industry of climate change abatement. The costs of reducing CFCs and SO_2 had proved to be much lower than industry's original estimate. The most important consideration was to protect the integrity of the Framework Convention by first instituting a limited trading scheme within Annex I which the developing countries could carefully scrutinize as a forerunner of an expanded system. Meanwhile, business needed to start taking a leadership role, moving forward to help governments set reduction targets in order to avoid the situation where governments were forced to impose measures business would dislike.

Michael Jenkins OBE, Chairman of the Futures and Options Association, expressed confidence in the transferability of many elements of the American SO_2 trading scheme to a global CO_2 emissions trading regime. He said sceptics would not be persuaded by intellectual arguments but only by a successful pilot phase.

David Porter, Head of the UK Association of Electricity Producers, said his members were convinced that some sort of emissions trading scheme was likely and that it would be preferable to taxes. However, consumers were just starting to take an interest in the climate change issue.

Dr Klaus Kabelitz, Head of the Economic and Energy Industry Department of Ruhrgas AG, commented that taxes and command and control measures would not get the job done. He explained how voluntary emission reduction commitments from its major industrial sectors were helping Germany meet its 1990/2005 25 per cent CO_2 emissions reduction commitment.

Arrangement of this volume

The next two chapters address the political context of climate change. Chapters 4 and 5 look at the evolution of and lessons from sulphur controls in the UNECE, while Chapters 6 and 7 consider the lessons from sulphur trading in the United States. The remaining chapters look at CO_2 controls: Chapters 8–10 consider the international context; Chapters 10–13 summarize national programmes and concerns about Joint Implementation; and the last three chapters discuss the options and prospects for tradeable emission permits.

Chapter 2

Global climate change: new policy directions in the US political context

Dirk Forrister

Introduction

In the environmental policy arena, climate change is a bigger question than most issues. It poses enormous environmental and economic challenges: the greatest number of constituencies in the greatest number of nations will feel the impact, whether by the actions that nations agree to take or by the forces of nature if nations do not agree.

As an Assistant Secretary of Energy with responsibility for congressional affairs, one of my primary responsibilities is advancing our administration's legislative and policy priorities through our political process. I bring a front-line perspective of where the thinking is in Congress on this issue, and I also have some depth of experience regarding the Clinton/Gore administration's work on climate policy. In this paper, I want to recap the main points of the US approach. I will analyse it from three vantage points – environment, economics and energy – all drawing upon the lessons we have learned from our acid rain law. Along the way, I will discuss some of the political calculus undergirding our framework.

Guiding principles in approaching climate change

From the outset of our work on climate change, our administration's thinking reflects basic principles of sustainable development: we pursue *environmental effectiveness, cost-effectiveness and energy security* at the same time. As a matter of principle, we do not accept the premise that the three concepts are in conflict. In fact, pursuit of environmental improvement can improve industrial efficiency, save money and improve energy security.

This policy approach sets a clear environmental objective in a solid target for improved performance, and it advances an economic objective by providing complete freedom of choice among compliance options so that solutions can be sought out competitively and seized at the least cost.

We expect this approach to advance our energy security as well. How? Primarily, we can promote improved fuel diversity, which for us means building additional resources from energy efficiency technology deployment, renewables, advanced fossil fuel technologies and advanced nuclear technologies. Next, since we prefer 'portfolio' approaches to energy – those that do not over-rely on single fuels or single sources of supply – we seek, as a policy matter, to advance the same 'portfolio' approach to environmental solutions. Just as we like competition among fuel options, we like competition among environmental solutions. This keeps costs down and innovations up.

These same approaches should *enhance trade competitiveness* as well, meaning that they make us better performers in international markets. Wastefulness is akin to being a fat and lazy competitor. We do not like to be wasteful, whether on energy or on environmental spending – by that I mean that we like to get the 'biggest bang for the buck' with every energy dollar and every environmental dollar.

These principles find support politically in the US from members of both political parties, industry, environmentalists and labour. The political context of our work on climate change is particularly interesting now. The environmental issue rose to prominence as one of the top four issues in the Clinton/Gore re-election campaign. Republicans struggled in numerous congressional campaigns to overcome public concerns that they had gone too far in considering anti-environmental measures. The end result is some measure of an electoral mandate in favour of the environment.

But we have yet to test just how much political running room is available to Democrats on climate change. The Republican attacks on the Environmental Protection Agency over the past two years put 'backyard' environmental issues back in the public eye. We Democrats scored political points just standing up for enforcement and retention of current laws. Climate change involves gaining new ground, new legislation. And public opinion is still ripening on the issue. Many industry opponents of climate action believe that, if we reach too far in adopting expensive climate proposals, the Republican Congress will get adequate political cover from moderate Democrats from the industrial belt and other fossil fuel producing or consuming states to enable them to block a ratification vote.

In our system, the Senate must ratify any protocol or amendment by a two-thirds vote. While Democrats won the presidency and gained some seats in the House of Representatives, both the House and Senate remain controlled by Republican majorities. And most importantly, the Republican margin grew in the Senate, so that they now hold 55 seats to the Democrats' 45. Even if we can hold every Democratic vote, we must win 22 Republican votes in order to prevail on ratification.

The key to gaining ratification is remaining true to the principles I just set forth: combined strength of environmental progress, continued economic growth, and energy security. In summer 1996, we began to enunciate a policy framework, born of these principles, that we believe has the greatest potential to bring parties together around a meaningful next step on this issue, with good prospects of political viability in the United States and, we hope, around the world.

New ideas for the framework

US Under Secretary of State Tim Wirth unveiled a policy proposal during the climate change negotiations in Geneva in July 1996, known as the Second Conference of the Parties to the Framework Convention on Climate Change. The basic tenets of this framework were reflected in the Geneva Declaration adopted by Ministers at COP-2. Although news accounts underscored the environmental policy news that we favoured a binding target, there was also good news on the energy and economic front that was not widely reported.

Wirth indicated no specific level of reduction – he essentially gave our views of 'how' the future regime should work, not 'what' or 'how much' should be done. This was the first time the US articulated our preferred approach in this detail. The highlights were:

We support realistic, verifiable and binding national targets

From the first year of our administration, we have maintained that the Rio Framework Convention on Climate Change was not adequate to meet its

ultimate objective. As our position matured, we focused on the shortcomings
associated with the non-binding nature of the Convention – countries make
grand statements about what they are going to do, and then fail to live up to
their promises. For the first time, the US stated that it preferred a binding target at a realistic level (i.e., that nations can afford to meet and do so in practice) using flexible national plans and international emissions trading to get
there. We also stated that we could not support the specific proposals tabled
so far by other nations, because they were too ambitious or inflexible.

We endorse maximum flexibility in pursuit of the target

We have long held strong views in favour of flexibility to achieve cost-
effectiveness. Our economic analysis, though still in progress, confirms
that we can achieve binding targets cost-effectively so long as we gain the
benefits of flexibility regarding where and when reductions are made.

On one point, I should stress our resolve: flexibility and a binding target
are firmly linked. We will not support one without the other. The only way
in which the US Senate is likely to agree to a binding greenhouse gas target is if there is broad flexibility to achieve the target at the least cost.

We offered several specific features of flexibility that would ensure
cost-effectiveness and mitigate against any harsh energy and economic
burdens of a binding target:

- *Timing* Phasing in a target more gradually permits industry more time
  to plan rationally and deploy technology effectively to meet the target.
  The US supports a binding system for the 'medium term' and not the
  'short term'. This means not 2005, the short-term date mentioned in the
  1995 Berlin Mandate, but more likely something closer to its references to 2010 and 2020.

  In environmental terms, more optimal timing and intelligent budgeting can provide greater benefits as well. Once built, large energy infrastructure projects are used for 50 to 75 years – in other words, for most
  of the century. In these instances, more reductions could be achieved
  overall by smart emission budgeting that allows 'super-efficient' technologies the time to mature rather than forcing early replacement of

central stations with only 'semi-super-efficient' technologies that are available in the near term. The difference in carbon loaded into the atmosphere during the period while technology is still maturing, for example from 2010 to 2020, could be dwarfed by the difference in the cumulative benefits that result in the decades that follow. The flexibility to budget over a multi-year target offers the promise of more climate benefit for less money.

- *Emissions trading and joint implementation* US support for a binding target fits hand-in-glove with support for maximum flexibility through joint implementation and international emissions trading. This is the first time that the US has so explicitly advanced an emissions trading policy model, leaving behind the other tools under consideration earlier in our administration, such as command-and-control regulation or energy/emissions taxes. Under this policy approach, the private sector is given freedom of choice in where to invest in mitigation and advanced technology deployment so as to achieve the greatest emission reduction for a given investment.

- *No mandatory policies and measures* The US cannot support mandatory, harmonized policies and measures. After months of discussion, we can see no possibility that a binding policy and measure approach will work, given the differences in policy preferences in various nations. To US industry so familiar with some of the rigorous standards in our country, 'harmonized measures' means an additional layer of international regulation – such as automobile standards, appliance standards or carbon/gasoline taxes, all of which they are certain would be defeated in the US Congress.

- *Long-term concentration goal* In addition to the medium-term binding target, the US supports working toward a long-term concentration goal (e.g., for the next 50–100 years). In all likelihood, this will require further negotiation beyond Kyoto when both developed and developing nations are negotiating on a global approach. The Framework Convention sets the ultimate objective of stabilizing atmospheric concentrations at non-dangerous levels, but most negotiations have focused on emissions rather than concentrations.

Since greenhouse gases remain active in the atmosphere for hundreds of years, annual emissions are not nearly as important as multi-year loading trends. For business, this is a tremendous concept, because it enables us to begin discussions of carbon budgeting over longer time-frames – as described above, we envision an approach where emissions rise for a number of years, followed by compensating reductions in later years. This would provide time to research, develop and deploy better technologies.

- *The future of voluntary programmes* Unfortunately, some observers are confused about where we stand on voluntary programmes. The press accounts and some industry reports misconstrued our move away from a non-binding international target and towards a legally binding target as also moving away from voluntary domestic programmes – and we have not reached decisions on domestic implementation. Our position is that the voluntary nature of the treaty has failed as various nations have missed their voluntary targets – but we did not mean to say that the voluntary programmes in our domestic plan had failed. They succeeded to the extent that Congress funded them, and some of them, such as the Climate Challenge with electric utilities, are tremendously successful with very little government funding.

We support coupling this approach with accelerated performance from developing nations

The climate problem is a global concern requiring global solutions. We have shown how we plan to take the lead in taking action, and we are eager to learn how our friends outside Annex I can advance their commitments. We will offer our proposals on specific levels and time-frames when our analytical effort provides a sufficient basis for judgment.

As we move forward, there is a great deal of analysis still under way regarding the specific levels and time-frames that we will be able to support.

US Under Secretary of Commerce for Economic Affairs, Ev Ehrlich, continues to oversee our economic analysis. The selection of models and bench-marking will be nearing completion towards the end of the year or early in 1997, after which we will be able to run specific scenarios that will enable us to make informed judgments about levels and time-frames. We expect the negotiations on these topics will not ripen until late spring or early summer 1997.

We will also continue to elaborate our domestic implementation policies. Our thinking is open on how to approach individual sectors. There is potential for building upon the voluntary programmes for sectors with demonstrated success (such as electric utilities) as we consider their contribution to the binding national target – an opportunity industry should use to its advantage by offering proposals for stronger voluntary commitments.

The framework's lineage: acid rain trading in the United State

The new US framework on climate change is a direct descendant of our acid rain control programme. It draws on our experience, although there are important differences. The basic characteristics of success in our acid rain programme are reflected in our new climate approach:

- *Solid environmental gain* Our Acid Rain Law set a cap to achieve a 10 million ton reduction in sulphur dioxide emissions. From the moment President Bush announced the level, it was a political 'done deal'. Rather than argue about the level, Members of Congress focused their efforts on getting a fair share of allowances for their constituent companies. As a political matter, the affected companies were transformed: they behaved as asset holders trying to increase the size of their assets, not as regulated entities manoeuvring for loop-holes.

- *Early banking incentive* In practice, the greatest environmental gain is one that nobody expected: most of the utilities are over-complying early and building reserves of allowances for sale or use later. We are achieving clean-up earlier, deeper and cheaper than ever expected.

- *Works better, costs less* My EPA colleague, Brian McLean, provides some stunning statistics about how few people are needed to run the acid rain programme and what a small budget is required (see Chapter 7). This bureaucratic saving is only the beginning: private-sector transaction costs are also lower than in other clean air programmes. More corporate money is available for actual clean-up – and hopefully there is ample remaining for job creation as a result.

- *Freedom of choice in compliance* In our acid rain trading programme, freedom of choice is the secret to success. Instead of regulatory requirements on technologies that can be used to meet an environmental standard, companies are told the total amount of sulphur dioxide they are allowed to emit across their entire operation. They can get to that level of performance in whatever way they wish. At the company level, this provides a valuable result: they can entertain offers from the scrubber salesman, who must compete against the natural gas co-firing salesman, who in turn must compete with the low-sulphur coal salesman, who must compete with the demand-side management salesman. The competition among compliance options drives the price down. And while the environmental benefit is guaranteed, this portfolio approach complements our *energy security*: there is free competition among the diverse array of fuels, so long as the environmental results are the same. In the end, the free choice *rewards innovation.* Companies have the incentive to adopt emerging technologies to reap the environmental rewards. In the old command-and-control regime, they would have waited for a bureaucrat to approve a list of permissible technologies, which caused delays and stifled innovation.
- *Political consensus achieved* After ten years of deadlock on acid rain, the cost-effectiveness of emissions trading provided the basis for forging a bipartisan consensus for action. While conservative members of both our parties detest command-and-control and new taxes, both applaud emissions trading as the market-based mechanism that they can accept. For the individual Member of Congress, emissions trading offered an interesting result: while heavy regulation and tax schemes turned them off, the 'horse trading' among formulae for allocating emissions allowances suited them well. From the environmental perspective, this approach had decided advantages: the currency was emissions – not tax revenues. If a tax solution had been considered, to the Ways and Means (tax) Committee, the prevailing question at the end of the day would almost certainly have been, 'How much money does it raise towards our revenue need?' In the environment and energy committees, the question on acid rain was, 'How can we allocate the emissions rights fairly without exceeding our target?' In the end, the

coalition that came together in support of acid rain legislation – a combination of moderate Republicans from environmentally progressive states with Democrats – will be the same coalition we will need to support climate action.

Lessons learned from acid rain in the United States

There are obvious parallels between climate and acid rain from the policy perspective:

- *No perfect analogy to climate* First, I want to stress that the parallels are not perfect. Acid rain policy has profound differences: a smaller number of sources, some different technological solutions, and differences in geographical impacts. But there are some basic tenets, some of which I have touched upon, that I want to highlight as common to our approach on both.
- *Binding target secures environmental gain* In the tradeable permits approach, the level of environmental protection is secured from the start, and the flexibility should never compromise the gain. This is our experience in acid rain, and it can prove true for climate change policy as well.
- *Keep system simple* The regulator's tendency is to provide for every contingency with a set rule. But in a trading regime, the cost savings potential is enhanced by simple rules and procedures. We have a bias against bureaucracy and bottlenecks in a trading programme. And in this regard, I can never be enthused about governmental clearing-houses or trading floors – this is the job of the private sector. It is perfectly suited for determining whether reductions are real and deserve purchase, and it knows how best to discount investments it considers to be more risky. This form of market discipline beats regulation every time. I even prefer, as a concept, competition among emissions trading floors rather than a monopoly for one.
- *Flexibility delivers technology and savings* In the US acid rain experience, industry is proving to be better at picking the right technologies and the most comprehensive strategies than regulators were in previous

regimes. The competition among compliance options discussed earlier is the key to this success.

Now I want to turn to a more focused look at how this flexibility can deliver benefits on climate change.

Target limited capital to the greatest good

After some recent international financial events such as the Mexican peso crisis, every nation – particularly developing nations – is sensitized to how connected world financial markets are now. When financial markets get constrained in the US, Europe or Japan, shock waves are likely to be felt through every economy in the world.

Keeping in mind this connectivity, we commissioned the Electric Power Research Institute (EPRI) to evaluate the cost savings attributable to flexibility. They looked at what the world economy would spend to achieve a 20 per cent reduction by 2020 and maintain that level for the century. Here is what they found:

- If reductions could be achieved only in the OECD, the cost would be $2.68 trillion.
- If those reductions could be achieved anywhere, the cost would be cut more than in half – to $1.16 trillion.
- And if the same reductions were achieved, but phased into optimal business cycles combined with geographical flexibility, the cost drops by $2.2 trillion to $0.51 trillion – over 80 per cent.

There is not a ton of difference in the environmental performance of these options, yet the $2.2 trillion savings achieved with optimal timing and geographical flexibility are roughly equivalent to one-third of the current annual economic output of the United States. And think of what the resources saved could accomplish worldwide for other national and international priorities.

Another way of viewing this analysis is to consider that most governments are likely to devote only a set amount of money to the climate problem – even if it comes from the private sector. Flexibility measures offer the promise that the fixed capital available will achieve the greatest environmental good. In other words, if governments decide only to spend $0.51 trillion, they could achieve a 20 per cent reduction. But without flexibility, at that price, the environmental gain would be less than a 5 per cent reduction.

This explains why some environmentalists are so strongly in favour of flexibility – it will ensure that the greatest environmental benefit is achieved with the fixed amount of capital available.

Global costs under four alternative flexibility cases

The US analytical effort further quantified the benefits of 'where' and 'when' flexibility. It set the combined benefit, similar to EPRI, at 85 per cent. If broken down, it quantified the savings from geographic flexibility to be 70 per cent, and timing flexibility to be 30 per cent. This makes clear the justification for our position favouring 'maximum flexibility'.

Mobilize private power finance

For many years, we have stressed the importance of tapping private-sector financing to turn it towards environmentally superior approaches to energy and industrial production. We have stressed that these markets are the key to our technology transfer priorities, since the private sector owns the technologies and has the financial resources to support deployment.

Just since the Rio accord was adopted, there have been some dramatic developments in international electricity markets. We are seeing growth as never before, and virtually all the action is in developing countries. Often, the international climate negotiations prompt comments about bilateral and multilateral aid funding. But the trends in this area are remarkable. Over the past ten years, there has been an explosion of financial investment in the electric power sector in developing nations. And the sources of this financial flow show an important trend:

- *1985* Of the roughly $17 billion spent in 1985, about one-third came from official bilateral and multilateral institutions. A small fraction came from private markets. Over half came from the local/national governments. The international private sector provided a small fraction – about 2 per cent.
- *1994* Ten years later, the total market had nearly tripled. The lion's share of the growth came from the international private sector – it grew by nearly 1000 per cent. Local/national public investment grew by over 50 per cent. Official bilateral/multilateral funding remained relatively flat.

I offer this to make a simple point: the key to the financing game is tapping the private financial flows. That is where the action is and will be. Again, private financing of energy and environmental projects leaves local and national governmental budgets more available for other domestic priorities. Our policy framework can help ensure that these private resources are used to finance the best technologies in the best locations for the climate.

Benefits of research and development: coal

Some of the 'best technologies' we hope can get financing will be clean coal technologies. For the foreseeable future, coal is going to continue to be a major energy player worldwide. As we look at this resource, it is important to note that technology can still bring tremendous benefits.

In a tradeable permits regime, coal can continue to thrive as a fuel, so long as it meets environmental requirements with technology or with compensating emission reductions achieved elsewhere. The US experience in the Department of Energy's clean coal technology partnership with industry is that we will cut carbon emissions from power plant designs by a third from 1980s standards by the year 2000. By 2015, we expect to be 47 per cent cleaner. We will build this effort from a strong track record: since the 1970s, we have improved coal's environmental performance on fly ash and sulphur dioxide by over 90 per cent, and nearly 80 per cent on nitrogen oxides. We are 'bullish' on our ability to continue these trends – and this time, we intend to do the same for carbon dioxide. And we intend to do so while reducing the cost for the power produced by 10–20 per cent.

Renewables: progress

We are working to ensure that renewable energy choices can compete in the emissions trading regime. Again, we should not overlook the dramatic progress made to date in improving performance and costs of renewables. The research and development efforts under way at the United States Department of Energy, in partnership with our private sector, have contributed towards progress on the cost and performance of the entire renewables portfolio:

- *Photovoltaic* Since 1980, the costs of photovoltaic electricity have improved by nearly 70 cents per kWh. By 2005, we are planning improvements to reduce the costs from just under 20 cents per kWh today to 12 cents per kWh – a 40 per cent decrease.
- *Wind* Since 1980, wind energy has improved from 40 cents to under 6 cents per kWh, and by 2005, we are charting a cost of only 3.5 cents per kWh – a 35 per cent decrease.
- *Biomass* As a benchmark on biomass, let me focus on one technology – ethanol: prices per gallon in the US have dropped from nearly $3.50 to $1.22, and we are charting a year 2000 cost of under $1.00 – a 20 per cent decrease.
- *Geothermal* Finally, geothermal energy is also making progress. In 1980, it cost around 9 cents per kWh, and is now in the 2 cent range. We expect it to drop to 1.2 cents by 2005 – a 30 per cent decrease.

The sum and substance of this technology pricing review is that costs are becoming more and more competitive. Our strategic investments in research and development in renewable technologies are beginning to pay off. Now we need the system in place to get the financing and policy drivers to connect the technology with the opportunities where the greatest emission reductions can be achieved. Our proposal for a tradeable permits system provides the policy signals that the market needs to see in order to make sound, strategic investments in renewables.

Emissions trading: basic needs

As we continue to elaborate our views on how to pursue a binding commitment with flexibility to achieve cost-effectiveness, there is a short list of basic requirements that will make the system work. It does not have to be complicated – at least, no more complicated than any approach to this mammoth issue.

Trading begins with a *baseline* from which the target is calculated and allocations are made. Reduction credits will be awarded for firms that can reduce emissions below the baseline. And firms will be required to compensate for exceeding the baseline with credits they have achieved or purchased.

In order to have confidence that credits are real, firms must provide *monitoring and tracking* of the reduction credits. Nations will need to *report and verify* that the reductions are not counted more than once or claimed by multiple firms. Nations will need to set *transparent, stable rules* to govern the system.

Finally, it will be important to a carbon trading system to develop a *budgeting* approach that allows nations and firms to *bank and borrow* pursuant to guidelines. Our experience on acid rain shows that banking provides an incentive for early action and over-compliance. And borrowing, within limits and with proper interest paid, can advance the interests in avoiding wastefulness and in delivering greater emission reductions in the long run. Borrowing should not be constructed to permit a series of intentional delays for free. Instead, it should ensure that, in the quest for greater cost-effectiveness, the environment gets more than just paid back – it gets an extra benefit through the interest paid. If carefully constructed, borrowing can be a win for the economy and for the environment.

Opportunities

Our experience with acid rain legislation teaches us that the tradeable permits model has decided advantages as an approach to climate change mitigation policy.

- No other approach delivers more environmental benefit for the limited funds available to nations.
- No other approach delivers more private-sector financing and technology to developing nations.
- No other approach delivers the best technologies to their greatest emission reduction opportunities.
- And no other approach delivers a higher likelihood of political success.

This is not to say that the tradeable permits approach is perfect. But as some wise sage once said of democracy: it may seem at times like the worst system of governance in the world – except for all the rest.

Chapter 3

Implementing sustainable development and CO_2 controls: initiatives from UNCTAD

Carlos Fortin

Introduction

The concerns which underlie the growing doubts over the environmental sustainability of contemporary production and consumption patterns have been factored into UNCTAD's work since the early 1990s. The Cartagena Commitment, adopted by UNCTAD VIII in 1992, was the first internationally agreed text to spell out the need systematically to re-examine concepts and tools of environmental analysis so as to incorporate the development dimension. Since then the concept of sustainable development has been integrated into all aspects of UNCTAD's work. We regard it as being primarily a development concept, as well as an environmental one. The challenge for developing countries is to eradicate poverty and increase employment and production through, *inter alia*, expanded trade while preserving and protecting the environment. Neither trade nor environmental preservation is an end in itself. The ultimate goal is the pursuit of sustainable development.

Also in 1992, UNCED recommended that UNCTAD 'should play an important role in the implementation of Agenda 21'. That challenge has been taken up. UNCTAD acts as task manager on trade, environment and sustainable development to the Commission on Sustainable Development, and we have since established a comprehensive work programme on trade and environment.

At the intergovernmental level, the interlinkages between trade, environment and development and related policies have been examined, in particular the effects of environmental policies, standards and regulations on market access and competitiveness; market opportunities flowing from the demand for 'environmentally friendly' products; and eco-labelling and eco-certification schemes.

In cooperation with other international and regional organizations, moreover, UNCTAD is undertaking a comprehensive technical cooperation programme on trade and environment designed to increase understanding of the complex issues involved, build institutional capacity in developing countries, provide information to policy-makers, and support the participation of developing countries in international deliberations. In this context, numerous policy-oriented studies have been carried out on areas such as eco-labelling and international trade and the effectiveness of different policy instruments in achieving the objectives of multilateral environmental agreements. Our Trade Control Measures Database has been adjusted to incorporate environmental measures which may have an impact on trade, and we are developing GREENTRADE, a computerized information system on environmental product concerns and measures. In November 1996, UNCTAD launched its BIOTRADE initiative, which is designed to help developing countries attain economic benefits from the full use of the Convention on Biological Diversity by increasing their capacities to compete in the emerging market for biological resources, as well as reducing transaction costs, increasing demand for biochemical resources, and enhancing conservation incentives.

This work has enabled us to identify some elements which are very relevant in developing strategies to make trade and environment mutually supportive and consistent with promoting sustainable development.

(1) There must be a full appreciation of, and sensitivity to, the differences in environmental priorities among countries. These differences highlight the importance of introducing political arrangements that provide adequate flexibility for local actors to resolve their own environmental problems, while remaining sensitive to the possible implications of national policies for international trade.

(2) Positing trade and environmental issues in the context of sustainable development requires taking account of some specific features of developing countries which complicate the policy options. Perhaps the most important one is the limited capacity of developing country governments to implement environmental measures owing to the inadequacy of available infrastructure, finance and technology.

(3) Solutions to environmental problems do not always lie only in environmental action as such. Certain policies which aim to promote economic development (such as improvements in infrastructure) also have positive environmental effects.

(4) Improved market access has an important role to play in the pursuit of sustainable development through providing resources for environmental improvements and increasing efficiency.

(5) Measures addressing transboundary and global environmental problems should, as far a possible, be based on international consensus. International externalities require regional or global agreement on norms and standards and raise the question of the distribution of the costs of environmental protection and improvement among countries. Thus, solutions to global problems need to be not only efficiency-based but also equity-based. This means that developing countries should receive transfers of funds and technology if their acceptance of efficient environmental solutions promises to place a heavy burden on them.

Attaining the efficiency objective is a complex undertaking. A major obstacle is the inability of markets to ensure on their own the environmental sustainability of economic activity. Markets often fail to recognize natural and environmental resources as assets, or to value properly the costs and benefits associated with the external effects of production and consumption activities. In such cases governments need to intervene in order to create the conditions for the internalization of the external costs associated with the resulting environmental damage. However, to avoid creating excessive rents, intervention should rely to the greatest extent possible on market-based economic instruments.

This thinking applies to the implications of global warming. The dynamics of global warming and the requirements of development appear to be on a collision course. Dealing with the first will require significant abatement of the emissions of greenhouse gases, in particular carbon dioxide. But throughout the developing world the need is for continued industrialization, which implies further increases in carbon dioxide emissions.

The way out of this conundrum cannot simply be to exempt developing countries from participating in international efforts at emissions abatement, although this might arguably be a reasonable temporary first step. Developing countries account for a growing proportion of global carbon emissions; and any international agreement that did not include provision for abatement by them would not be able to achieve its objectives over time. It also needs to be underscored that developing countries themselves have a strong interest in global abatement and in the methods of achieving it. All of them would be affected by climate change and, for some, the consequences of drought or a rise in the sea level could be calamitous.

Rather, the solution must be to seek policy approaches that meet some basic criteria. First, they should be capable of bringing about the overall global abatement in emissions required by the collective judgment of the international community; second, they should allow continued economic development in all countries, and enable the process of industrialization to proceed unimpeded in those countries in which this process is incomplete; and, third, they should provide appropriate incentives to ensure that industrialization in developing countries, and the replacement, renewal and expansion of industrial facilities in developed and developing countries, embody technology and practices that are emission-efficient.

Drawing inspiration from these criteria, one important avenue through which UNCTAD has sought to promote sustainable development has been the use of innovative, market-based mechanisms which have the capacity to combine cost-effective reductions in pollution with the generation of financial resources. In 1990, the UNCTAD secretariat established a research and development programme to examine the design and implementation of a global greenhouse gas emissions trading system to provide governments with an alternative policy mechanism to support the implementation of the emerging Framework Convention on Climate Change. Since then, five major reports and two illustrated syntheses have been published on the subject. In December 1996 we released the results of our most recent work, in two volumes, one on the legal and institutional and one on the organizational aspects of establishing a pilot international greenhouse gas trading system.

The report on legal and institutional issues represents a major advance in our understanding of matters essential to the early establishment of a pilot emissions trading programme. The near-term prospects for international emissions trading have brightened significantly following the adoption in July 1996 of the Geneva Declaration by the second Conference of the Parties to the Framework Convention on Climate Change. As you know, negotiations on a new protocol to the Framework Convention will now focus on agreeing 'quantified legally binding objectives for emission limitations and significant overall reductions within specified time frames'.

UNCTAD's work on the development of a global greenhouse gas trading system is therefore underpinned by three key assumptions. First, climate change is a global problem that will affect developed and developing countries alike. All countries will therefore need to contribute to its solution, though the developed countries must bear the primary responsibility for these efforts: not only are they mainly responsible for global warming, but they are also best placed to bear the burden associated with finding solutions to it. Second, climate change raises fundamental issues of industrialization and development. These processes remain incomplete in developing countries. Global cooperative solutions to climate change must recognize the rights of developing countries to pursue their industrialization and sustainable development, and by implication, increased emissions of greenhouse gases, especially carbon dioxide, by those countries for the foreseeable future. Third, a global greenhouse gas emissions trading system would leave each country free to choose its own domestic policy mix for controlling greenhouse gases emissions at the national level (emission taxes, charges, regulations, etc.).

With the successful development of the sulphur dioxide trading programme in the United States, it is now widely recognized that trading systems are efficient means of controlling emissions of greenhouse gases at minimum cost. Emissions trading provides both efficiency and flexibility in meeting established targets and timetables, and dynamic incentives for the development and application of new technologies.

UNCTAD believes that the time has come to advance further towards the development of a pilot greenhouse gas trading system. We see this approach as being firmly rooted in the Framework Convention on Climate Change; as being consistent with the expected provisions of the protocol to the Framework Convention; and as having a voluntary nature. A gradual, step-by-step approach to the construction of a greenhouse gas trading system is indispensable if we are to minimize the negative consequences of what is obviously a complex process of market innovation. A pilot programme would facilitate the process of 'learning by doing', ensuring that the new system is underpinned by empirical research and development activities. This would help to build confidence in it.

We are convinced that developing countries and countries in transition should participate in an international greenhouse gas trading system. Their participation would help to lower the cost of abatement worldwide, making abatement economically feasible and increasing the political acceptability of deeper reductions. These countries would benefit financially and technologically from the flow of direct investments into emission reduction and sequestration projects (in the framework of Joint Implementation/Activities Implemented Jointly), as well as from the flow of finance and the transfer of technology that would be generated via sales of emission allowances and credits. And they would, of course, also share in the benefits of effective global abatement.

Our own estimates of potential resource transfers to recipient countries are in the region of US$40–50 billion annually – roughly equivalent to the current level of official development assistance flows to developing countries. Clearly, emissions trading could significantly influence the pace and direction of the evolution of the international trade and financial system. This, of course, is a critical issue. Therefore, while we recognize that legally binding emission targets and timetables should be applicable only to Annex 1 Parties (developed countries), trading systems could and should be designed to include other parties under mutually agreed conditions.

With these considerations in mind, we have been encouraging public/private partnerships in order to implement a pilot greenhouse gas trading programme. We are currently working with the Earth Council and Centre Financial Products to develop a pilot emissions market through the establishment of the Global Environmental Trading System (GETS). We see the GETS initiative as an important extension of our own efforts and we are fully committed to its success.

This conference is taking place at a very timely juncture. Not only has climate change taken on overwhelming political significance with the clarification of policy by one of the major players in the negotiations on a protocol to the Framework Convention, and the rapidly approaching deadline for the conclusion of these negotiations in 1997, but 1997 also marks a significant milestone in the history of multilateral cooperation on international environmental problems. Very soon the international community will begin assessing the progress made in the implementation of Agenda 21, five years after the Rio Earth Summit. That unprecedented gathering of world leaders had brought great hope and raised expectations. While it is evident that some progress has been achieved, for example in controlling substances that deplete the ozone layer, progress in other critical areas has fallen short of expectations. Infusing new momentum and enthusiasm into the process of consensus-building and action will not be easy. Innovative ideas are needed. More forceful and decisive leadership on environmental issues is also essential if we are to make the transition from traditional resource-based economic development to environmental sustainability. This conference is part of a necessary process of reflection and assessment, and can make a lasting contribution to charting the way forward.

Chapter 4

The influence of science on the evolution of international regimes on sulphur

Harald Dovland

Introduction

Towards the end of the 1960s, Swedish scientists presented data showing that an area of acid precipitation over central Europe had been expanding from year to year to also include southern parts of Scandinavia. It was argued that this was caused by increasing emissions of sulphur dioxide from the burning of fossil fuels in Europe. At the same time it was noted that lakes and rivers in southern Scandinavia had become more acid and that fish stocks had declined and even disappeared in acidified freshwater systems. Scientists argued that there was a link between increasing sulphur emissions, increasing acidity in precipitation and freshwater, and the disappearance of fish stocks. The Nordic countries soon claimed that the acidification was an international air pollution problem and that pollution control might be necessary in other countries than where damage occurred. The issue was raised at meetings of international organizations, but there was far from general agreement about this issue. It was not accepted that air pollutants could be transported several hundreds of kilometres in the atmosphere and then cause environmental damage.

Agreement was, however, reached on an international study within the OECD (OECD Programme on Long-Range Transport of Air Pollutants, 1972–77). Through this and several other national and international research and monitoring programmes, it became well established that air pollutants were transported over long distances (>1000 km) and that pollutants emitted in one country could cause environmental damage in another. The best-documented damage is the acidification and loss of biological life (including fish) in lakes and rivers in southern Scandinavia. This is mainly caused by the deposition of sulphur and nitrogen oxides originating elsewhere in Europe. In the 1980s, widespread forest damage was reported from many European countries. The cause of the damage is not yet fully

established, but there is general agreement that air pollutants such as acidi-
fying substances and photochemical oxidants play an important role.

Policy responses to scientific findings

The first clear international policy response to these scientific findings was
the adoption in 1979 of the Convention on Long-Range Transboundary Air
Pollution under the United Nations Economic Commission for Europe
(UNECE), which includes all European countries as well as Canada and
the United States. According to the basic obligation of this Convention, the
Parties 'shall endeavour to limit and, as far as possible, gradually reduce
and prevent air pollution including long-range transboundary air pollution'. This is not a strong policy response; it is hardly more than a statement of intent, partly owing to lack of political awareness at that time and
partly because of scientific uncertainties.

The situation changed at the beginning of the 1980s. Not only did policy-
makers begin to realize that large parts of Europe were affected by transboundary air pollution, but they also had to recognize that more than 'acid
rain' was involved. A dramatic change in attitude was brought about by the
observed forest damage in central Europe, and the Nordic countries were
now joined by a number of other countries ready to act nationally and internationally to control air pollution. The change was, however, caused more
by growing public and political awareness than by new scientific findings.

An important step in the implementation of the Convention was the adoption in 1985 of the Protocol on the Reduction of Sulphur Emissions or their
Transboundary Fluxes by at least 30 per cent. For the first time, agreement
had been reached on specific international obligations. A few countries with
large sulphur emissions, including the United Kingdom and Poland, did not
sign the protocol. Thus, it took more than 15 years from the time the problem was brought to the attention of international organizations until a specific agreement was signed. The goal of reducing national sulphur emissions
by 30 per cent was, however, based more on political realities than on scientific results. Science had shown that larger emission reductions were needed.

The Sulphur Protocol was later followed by the 1989 Protocol on the
Control of Nitrogen Oxides (basic obligation to freeze NO_x emissions at

the 1987 levels before 1995) and the 1991 Protocol on the Control of Emissions of Volatile Organic Compounds (VOCs) (30 per cent reduction of the 1988 levels by 1999). Thus, the 'first generation' of protocols was based on equal percentage emission reductions for all countries involved, generally without any evaluation of whether this was a cost-effective way of improving the environment. The VOC Protocol, however, provides possibilities for differentiated reductions by allowing countries to include only those parts of their territories where emissions contribute significantly to transboundary fluxes.

In the second half of the 1980s, scientists initiated work which could lead to a 'new generation' of protocols based more on cost-effectiveness and environmental needs than on politically acceptable percentages. This is the 'critical load approach' (explained below), which was the basis for the Second Sulphur Protocol, signed in Oslo in 1994, and for the present negotiations on a multi-pollutant (NO_x, VOCs and ammonia) protocol. 'Critical load' has been defined as a quantitative estimate of an exposure to one or more pollutants below which significant harmful effects on specified sensitive elements of the environment do not occur according to present knowledge.

In the negotiations for the Second Sulphur Protocol, results from integrated assessment models were taken as the starting point for agreeing on differentiated emission reductions for individual countries. The criterion for optimization was the '60 per cent gap closure', and the calculations showed how this could be achieved at the least cost for Europe. An element of flexibility was introduced in the negotiations by allowing for different target years (2000, 2005, 2010). Table 4.1 illustrates the variations among some countries with respect to emission reduction commitments, here expressed as percentage reductions relative to the emission level in 1980.

To increase the cost-efficiency of the agreement further, the Second Sulphur Protocol includes the possibility for countries to implement commitments jointly. However, the rules and procedures for joint implementation have not yet been agreed.

Although both European and North American countries are parties to the Convention on Long-Range Transboundary Air Pollution, its importance has probably been greatest for Europe. Moreover, Canada and the

Table 4.1: Percentage emission reduction commitments (reduction from 1980 level, %)

	2000	2005	2010
Austria	80		
Czech Republic	50	60	72
Denmark	80		
France	74	77	78
Germany	83	87	
Hungary	45	50	60
Norway	76		
Spain	35		
Sweden	80		
United Kingdom	50	70	80

Source: Second Sulphur Protocol, Oslo, 1994 (UNECE, Vienna).

United States have undertaken scientific studies and monitoring programmes which have led to agreements on emission reductions to diminish acidification, but without using the critical loads approach to develop cost-effective emission scenarios. The United States has chosen emission trading as a main instrument to reduce the abatement costs. The US experiences with emission trading show that emissions reductions have been achieved at significantly lower costs than estimated when the legislation was prepared.

In Southeast Asia, programmes to monitor acid rain have been initiated. So far no significant damage has been documented, but emission projections show that the critical loads are likely to be substantially exceeded in a couple of decades. A challenge for that region is therefore to make use of the scientific information available to prevent 'exceedance' of the critical loads and then also avoid environmental damage.

Critical loads approach

In theory, the best way to solve an environmental problem caused by transboundary air pollution would be to take as a starting point the level of deposition or concentration of air pollution that an ecosystem can tolerate without negative effects. On the basis of the 'critical loads' or 'critical levels', which will often vary geographically, an optimal strategy for reduction of emissions can be elaborated using information on atmospheric processes and cost-efficiency of available control technologies. To use the critical load approach, the following information is needed:

(1) Inventories of current emissions and projections of future emissions.
(2) Estimates of potential for and costs of emission reductions.
(3) Long-range transport models describing in quantitative terms the transport of air pollutants within the region.
(4) Maps of critical loads.
(5) Integrated assessment models in which the above elements are combined to assess alternative reduction strategies.

To provide the substantial amount of information needed on each of these elements, the cooperation under the Convention has made use of several Working Groups, Task Forces, various coordination centres and many research institutes. The following sections set out a number of comments on the elements of the critical load approach.

Inventories of current emissions and projections of future emission rates
Parties to the Convention report annually on national emissions of the relevant pollutants according to agreed guidelines, as well as emissions in grid-elements for model calculations. The quality of the emission figures varies from one country to another, but is in general improving. Emission data for sulphur dioxide are most reliable, while for nitrogen oxides, and in particular for VOCs, they are less satisfactory.

Emission projections are very sensitive to key assumptions about economic growth, energy use, technology development, etc. Work is under way to harmonize assumptions made for running projection models. For sulphur dioxide emissions, which originate mainly from power plants, projections were based on energy scenarios available through various international organizations.

Estimates of the potential for and costs of emission reductions

Considerable operational experience is available on techniques to reduce emissions of sulphur and nitrogen oxides, especially in the case of large combustion installations. The costs of applying the different control options can therefore be estimated on the basis of international operating experience of pollution control equipment gained in Europe during recent years. Where necessary they may be adapted to country-specific conditions of its application (local fuel qualities, boiler sizes, capacity utilization, etc.).

By ranking the available pollution control options according to their costs, 'national cost curves' can be established describing the cost-efficient combination of control measures to achieve specified levels of national emission reductions. An international comparison of these 'cost curves' shows significant differences in abatement costs among countries, reflecting, for instance, the difference in structures of national energy systems. The uncertainty of cost curves has not yet been fully evaluated.

At present, calculations take into account 'end of pipe' technology and a limited number of other ways to reduce emissions (e.g. combustion modification), while structural changes in the energy system, such as conservation or fuel substitution, have not yet been included in the analysis. Some preliminary analyses have, however, indicated that structural changes may contribute significantly to reduced costs.

Long-range transport models

A central tool in the critical load approach is the long-range transport models used to estimate the deposition and concentration of air pollutants resulting from a particular emission field. Development of such models has been a core activity within the European Monitoring and Evaluation Programme (EMEP) since it was initiated in 1977. These models give transfer matrices of atmospheric long-range transport over Europe for sulphur dioxide, nitrogen oxides and ammonia. Model calculations of ozone formation in Europe are also available, but owing to non-linearities, it is not a straightforward matter to provide transfer matrices in this case. The model results are regularly compared with measurements of air and precipitation quality undertaken within EMEP.

The present EMEP models provide results with a spatial resolution of 150x150km, but work has been started to improve the resolution to 50x50km.

Maps of critical loads and levels

On the basis of agreed methodology, Parties to the Convention have cooperated in preparing maps of critical loads for total acidity, sulphur and nitrogen deposition. For those countries that have not submitted any critical load data, a coordination centre has prepared maps based on available information, e.g. on soil characteristics. Although work will certainly continue to improve the data, sufficient information is available for application in abatement strategy development.

Integrated assessment models

Several models exist combining the different elements of the critical load approach, and these can be used as tools to assess alternative emission reduction strategies. The one most widely used is the Regional Acidification Information and Simulation Model (RAINS), developed at the International Institute for Applied Systems Analysis (IIASA).

Results from integrated assessment models show that to reach critical loads everywhere in Europe will require very large emission reductions – up to 100 per cent. It is hardly realistic to assume that such large reductions will be achieved everywhere in Europe. Various alternatives to formulate intermediate steps are therefore essential. For the Second Sulphur Protocol a reduction of the difference between present depositions of sulphur and the critical loads by at least 60 per cent was used as the criterion for optimization (often referred to as the '60 per cent gap closure').

Concluding remarks

Experience with the role of science in the 'acid rain negotiations' indicates that policy-makers' willingness to accept and make use of scientific findings depends on several factors apart from the quality of the science. It took rather a long time to agree on the Convention, while the critical load approach was accepted fairly quickly. Science had certainly provided more knowledge in the meantime, but other factors were perhaps even more important. At the time when the Convention was negotiated, the main effect of acid rain that had been documented was freshwater acidification in southern Scandinavia (and even that was questioned by some scientists). That was probably regarded as a 'remote' problem by most central European governments. When forest damage reports started to appear at the beginning of the 1980s, the situation changed rapidly. First of all, many governments had to acknowledge environmental damage in their own countries, and thus the environmental effects were more noticeable. At the same time, growing awareness and concern among the public had to be taken seriously. The situation in the mid-1980s was therefore very different from that of ten years before, and policy-makers were perhaps more receptive to scientific findings because of pressure from the society to solve an environmental problem. When the problem exists only in a neighbouring country, the focus is perhaps more on scientific uncertainties.

Although there seems to be an increasing willingness to make use of science in the efforts to reduce effects of acid rain in Europe, the first generation of protocols, with an equal percentage emission reduction for all countries involved, is based more on political considerations than on scientific input, except for the scientific fact that emission reductions are needed to solve the acid rain problem. The critical load approach, however, offers a procedure for developing optimized abatement strategies on the basis of the effects of air pollutants, making full use of our scientific understanding of the problem.

There are, of course, uncertainties associated with the various elements of the critical load approach. In addition, meteorological variability and the fact that so far only national emissions are regulated impose some restrictions on the further refinement of the critical loads approach. It is therefore hardly realistic to expect that commitments in future protocols can be copied directly from the computer output. But the scientific results provide a very useful starting point for negotiations.

The present negotiations on a multi-effect, multi-pollutant protocol aiming at emission reductions of nitrogen oxides, ammonia, VOCs, and possibly also sulphur dioxide, pose additional challenges both to scientists and to policy-makers. However, by considering more gases simultaneously, cost-efficiency may be significantly improved.

Chapter 5

Transboundary initiatives for controlling sulphur and possible lessons for CO_2

Tim Jackson and Peter Bailey

Introduction

The concept of joint implementation (JI) – or activities implemented jointly (AIJ) – now appears either implicitly or explicitly in the language or within the policy frameworks of a number of international conventions. These include the Framework Convention on Climate Change (FCCC), the Biodiversity Convention, the Montreal Protocol (for controlling the release of chlorofluorocarbons (CFCs)) and the Convention on Long-Range Transboundary Air Pollution (CLRTAP).

The general intention underlying each of these initiatives is to devise mechanisms which allow two or more Parties to meet their obligations under the conventions through activities implemented jointly. For instance, it has been envisaged that one Party to the FCCC (the 'donor') might invest in greenhouse gas emission abatement technologies within the geographical borders of a second Party (the 'host'). Ultimately, it has been suggested, appropriate institutional arrangements might be put in place to allow the donor to seek full or partial credit towards its own obligations under the convention for emission reductions resulting from these investments in the host country. The main argument for such a procedure has been on the grounds of economic efficiency: the costs of greenhouse gas abatement might be lowered by seeking out the least-cost options first, irrespective of geographical boundaries.

In spite of general agreement that joint implementation offers potential benefits in meeting international environmental targets, it should be pointed out that there has been some political opposition to the concept. This opposition has arisen largely from a perception that JI might turn out to be a way in which developed countries (as donors) could avoid taking action at home by intervening in developing countries (as hosts) in ways disadvantageous (in the longer term) to the latter. Whether or not such fears turn out

to be justified will depend crucially on the particular institutional and economic frameworks under which joint implementation arrangements are
implemented. As yet, these frameworks have not been specified under any
of the conventions; and it is prudent to recognize that there are a number
of methodological difficulties involved in devising such frameworks.[1]

This chapter takes a brief look at some of the issues which contribute to
those difficulties. We examine in particular attempts to devise joint implementation arrangements in the context of sulphur emissions, and discuss
the relevance of these issues for the case of greenhouse gas emissions. The
work described here is based on a project supported by DG-XII under the
Framework IV programme. That project is summarized briefly in the following section.

Accounting and accreditation of activities implemented jointly

Among the difficulties associated with using joint implementation as an
international policy measure are those associated with accounting for
emission reductions, accounting for the costs of those reductions, and
devising a mechanism for appropriately assigning credits for the reductions between donor and host countries. These issues are the subject of a
major international study involving six separate European institutions.[2]
The overall objective of that project is:

● to examine the concept of joint implementation as an institutional
  instrument relevant to the efficient and fair abatement of greenhouse
  gas and sulphur emissions.

Specific objectives are:

● to gather information and data relating to 'pilot phase' joint implementation projects set up between north and east European countries in the
  context of the FCCC;

[1] Jackson, T., 'Joint Implementation and cost-effectiveness under the Framework
Convention on Climate Change', *Energy Policy* 23(2), 1995, pp. 117–38.
[2] University of Surrey, University of York, Stockholm Environment Institute, Risø
National Laboratory (Denmark), Nutek (Sweden), and EVA (Austria).

- to gather information and data relevant to the operation of joint implementation arrangements between north and east European countries
  under the Oslo Protocol of the CLRTAP;
- to carry out a broad-based analysis of the data collected from these
  pilot projects under the following criteria: in technical terms, in economic terms, in terms of environmental effectiveness and in terms of
  distributional and institutional consequences;
- to examine the methodological basis on which it might be appropriate
  to account for emission reductions from joint implementation projects;
- to examine the methodological basis on which it might be appropriate
  to assign credit to donor nations for joint implementation projects carried out in host nations;
- to explore the use and relevance of integrated assessment models in
  evaluating joint implementation projects;
- to report on these examinations with a view to informing EU policy
  and national policies within the European Union in relation to joint
  implementation arrangements under the FCCC and the CLRTAP.

The project commenced in June 1996 and will report to the Commission
before the end of 1998.

Sulphur emissions and the Oslo Protocol

The 1994 Oslo Protocol is one of several protocols to the 1979 Convention
on Long-Range Transboundary Air Pollution organized by the United
Nations Economic Commission for Europe in Geneva. It is the follow-up
to the First Sulphur Protocol, the 1985 Helsinki Protocol. Thus the Oslo
Protocol is sometimes called the Second Sulphur Protocol (SSP).

Sulphur emissions have a wide range of environmental impacts at different spatial scales. However, the Oslo Protocol is primarily concerned
with acidic deposition. It was implicitly assumed that building damage or
human health effects were either local, i.e., not transboundary issues, or
would be improved anyway as a result of the emission reductions contained in the protocol. The Oslo Protocol incorporated the 'critical load'
concept, which is a measure of the natural environment's sensitivity to

acidic deposition and varies by location. Broadly speaking, the critical load refers to the amount of acidic deposition which can fall on a given area before harmful environmental impacts occur.

The relationships between emissions and depositions are calculated by the EMEP programme as part of the CLRTAP. Emissions are accumulated within national geographical boundaries. Deposition d_j is mapped onto each 150x150 kilometre square according to the formula:

$$(1) \qquad d_j \ = \ b_j + \Sigma_i \, e_i . t_{ij}$$

where b_j is the background deposition on the jth square, e_i is the emission from the ith country and t_{ij} is the 'transfer coefficient' from the ith country to the jth square.

For the most part, actual deposition levels in Europe considerably exceed critical loads. The primary environmental objective of the Oslo Protocol was therefore to reduce sulphur deposition below critical loads,[3] and the substance of the protocol was to set national emission ceilings – i.e., limits to e_j for each country j – with the declared aim of reducing the gap (the 'exceedance') between existing deposition levels and agreed critical loads.[4] For example, these deliberations resulted in emission ceilings for the UK which require a 50 per cent reduction in sulphur emissions by the year 2000, 70 per cent by the year 2005 and 80 per cent by the year 2010, relative to 1980 emission levels.

Sulphur emissions trading

European regulators have predominately relied on non-market policy instruments to reduce the levels of sulphur dioxide emissions into the atmosphere. The most common regulatory instruments in Europe are emission standards for large point sources, fuel quality standards such as maximum sulphur contents in oil products, and national emission ceilings such

---

[3] More correctly the critical sulphur deposition levels shown on the map in Annex I of the Protocol.

[4] These national emission ceilings were negotiated on the basis of the so-called 'gap 60' scenario in which exceedance of critical loads was to be reduced by 60 per cent.

as those set down by the Oslo Protocol. However, in recent years, market instruments for sulphur emission control have been assessed both by economists and by regulators who are keen to implement reduction strategies in more cost-effective ways. The most favoured instrument in Europe has been emission charges (or taxes); several Scandinavian countries and a number of central and east European countries have introduced charges on the emissions of a number of air pollutants including sulphur dioxide.[5]

Trading in sulphur emission permits or emission reduction credits has not occurred in Europe so far, although some 'bubble' limits have been set up that give a degree of flexibility on how sources can meet an overall bubble limit. This is in contrast to the situation in the United States, where emission trading has been used for many years and is one of the main policy instruments employed to reduce emissions of sulphur dioxide in the Clean Air Act Amendments of 1990.

The principal difficulty faced in designing sulphur emissions trading schemes is the spatial nature of the pollution. Emissions of sulphur from one country are not equivalent to emissions from another country because they will be deposited in different geographical areas with potentially diverse tolerances for acid deposition. In mathematical language, the transfer coefficients t_{ij} are not identical for all countries i, for all squares j. Thus a unit of sulphur emission abatement in one country cannot be equivalent to a unit of sulphur abatement in another country, and one-to-one trading of emissions is generally invalid.

It is worth noting here that the sulphur trading system introduced in the US Clean Air Act Amendments *does* have a one-for-one trading system. Effectively, the US system simply ignores the spatial characteristics of sulphur pollution. Or more correctly, we should say that the compromise of allowing one-to-one trading was felt to be worth the gains in simplicity. In practice, there is a range of safeguards in the US system such as state-level restrictions on ambient air quality, and simulation modelling work of likely future trades has given some indications that a 'bad' spa-

[5]Swedish Ministry of the Environment, *The Swedish Experience - Taxes and Charges in Environmental Policy*, Swedish Ministry of the Environment and Natural Resources, Stockholm, 1994.

tial pattern of emissions is unlikely to occur. In Europe, these compromises have proved both politically and scientifically more problematic.

Joint implementation under the Oslo Protocol

During preparatory work for the 1994 Oslo Protocol, European systems of sulphur emissions trading were examined by the Task Force on Economic Aspects of Abatement Strategies and by the Working Group of Strategies – the main negotiating body of the CLRTAP. The Protocol itself contains 'enabling' language, designed to allow and perhaps further the development of joint implementation within the Convention. Article 2, paragraph 7 states:

> The Parties to this Protocol may, at a session of the Executive Body, in accordance with rules and conditions which the Executive Body shall elaborate and adopt, decide whether two or more Parties may jointly implement the obligations set out in annex II. These rules and conditions shall ensure the fulfilment of the obligations set out in paragraph 2 above and promote the achievement of the environmental objectives set out in paragraph 1 above.[6]

At present the rules for joint implementation of sulphur emission reductions are the subject of ongoing negotiation. It is likely that a certain number of institutional conditions will be imposed:

- Only Parties to the Protocol may enter into a joint implementation agreement.
- The proposal shall specify the part of its emission reduction obligation one Party will implement through reductions by another Party.
- The proposal shall specify the emission reduction the other Party will undertake in addition to its obligation in accordance with the Protocol.
- The proposal shall contain an assessment of the deposition impact.

---

[6] UNECE, *Protocol to the 1979 Convention on Long-Range Transboundary Air Pollution on Further Reduction of Sulphur Emissions*, United Nations Economic Commission for Europe, Geneva, 1994.

- The proposal shall indicate the level of expected cost savings and the means of compensation chosen.

For reasons already discussed, difficulty has arisen on how to devise a system of joint implementation that helps Parties to the Protocol meet their emissions reduction targets (the obligations in annex II and paragraph 2) in a cost-effective manner that is also consistent with the critical load concept (the environmental objectives of paragraph 1). There is general agreement on how emissions cause a change in deposition (see equation 1 above) and why sulphur emissions cannot be traded on a simple one-to-one basis in Europe.

It has been argued that it is possible to convert sulphur emissions into a quantity that *can* be traded by using a sulphur emission 'exchange rate' r defined according to a formula of the following kind:

$$(2) \qquad \Delta e_D = -r_{DH} \cdot \Delta e_H$$

i.e., the allowable increase in emissions (Δe_D) above the target level in the donor country D is some fraction r_{DH} of the decrease in emissions (Δe_H) below the target level achieved in the host country H. The exchange rate r is supposed to capture the spatial differences which characterize deposition as a result of emissions from the donor country as compared with emissions from the host country.

A little reflection reveals that exchange rates, at least as characterized by equation 2, offer an over-simplification of a potentially very complex situation. In reality, although it is certainly possible to calculate an effective exchange rate r from each transaction, these rates are far from simple, uniform parameters which can be determined on a once-for-all basis. In general, the value of the exchange rate will depend crucially on a number of other potential constraints which might be imposed upon the situation.[7]

[7] Klaassen, G., *Trading Sulphur Emission Reduction Commitments in Europe: A Theoretical and Empirical Analysis*, International Institute for Applied Systems Analysis, Laxenburg, 1995;. Førsund, F.R., *Sulphur Emission Trading*, Working Paper, Department of Economics, University of Oslo, 1993; Bailey, P.D., Gough, G.A., Millock, K. and Chadwick M.J., 'Prospects for the joint implementation of sulphur emission reductions in Europe', *Energy Policy* 24(6), 1996, pp. 507–16.

For instance, it is likely that little or no increase in deposition will be allowed as a result of a joint implementation agreement in areas where exceedance already occurs. In general, each potential transaction is likely to require some form of environmental impact assessment to check that the resulting deposition levels do not contravene a specified set of rules and conditions. Present discussions suggest that these might include (at least) the following set of constraints on any joint implementation agreement:

(1) it shall not lead to an increase in European exceedance of critical loads;
(2) it shall not increase deposition in any EMEP square by more than a specified percentage; and
(3) it shall not increase deposition in third-party regions by more than a specified percentage.

From these rules alone it is possible to conclude that a sulphur exchange rate (as defined by equation 2) is quite likely to be less than one, and that we cannot expect to find simple transitive or inverse relations between exchange rates.[8] More importantly, as the following section reveals, we cannot assume that the exchange rate remains constant between two countries throughout all transactions.

Sulphur exchange rates – a simulation

We present here the results of modelling a hypothetical sulphur trading relationship between Sweden and Estonia. This example has been chosen only for illustrative purposes – and in spite of the fact that Estonia is not at present a Party to the Oslo Protocol.[9] Let us assume that Estonia does become a Party to the Protocol and that Sweden wishes to reduce its requirements under the Protocol for emission abatement in 2010 by investing in emission abatement in Estonia. Let us further assume the following conditions:

---

[8]   For example, it is likely to be incorrect to assume the simple inverse relation $r_{HD} = 1/r_{DH}$ which equation 2 seems to imply.
[9]   On the other hand, it is informed by a continuing Swedish interest in bilateral investment in emission abatement in the Baltic states.

- the transaction must reduce, or maintain at the same level, the total exceedance of critical loads in Europe; and
- the transaction must not increase deposition in any EMEP grid square by more than a pre-specified percentage (y%).

The critical question we address in this simulation is the following: by how much would it be allowable for Sweden to increase its sulphur emissions (above 2010 target levels) for each kilotonne (kT) of additional emission abatement in Estonia?

This question was examined using the SEI CASM acidification model for a range of host country emission abatements from 10 to 100 kT.[10] Using as a reference point the emissions in 2010 that would have resulted from the implementation of the Oslo Protocol, we have calculated the maximum allowable emission increases in Sweden for varying grid level constraints – from no allowable increase in grid level deposition to a 5 per cent grid level increase.

A couple of points are worth making in relation to the maximum allowable donor country emission increases and the sulphur emission exchange rates.

- First, it is very clear from this analysis that sulphur exchange rates are highly dependent on the actual rules and constraints under which joint implementation takes place. Differing grid level constraints impose differing exchange rates.
- Secondly, it is to be noted that the allowable increase above target levels in the donor country is less than the reduction below target levels in the host country. Even with no constraint imposed at the grid level, the rule of no increase in exceedance (rule 1 above) still prevents donor country emission increases from matching the host country decreases in emissions. For cases where constraints are imposed at the grid level,

---

[10] Bailey, P.D., Gough, G.A., Millock, K. and Chadwick, M.J., 'Prospects for the joint implementation of sulphur emission reductions in Europe', *Energy Policy* 24(6), 1996, pp. 507–16; Gough, G.A., Bailey, P.D., Biewald, B., Kuylenstierna, J.C.I. and Chadwick, M.J., 'Environmentally targetted objectives for reducing acidification in Europe', *Energy Policy* 22(12), 1994, pp. 1055–66.

allowable donor country increases are considerably lower than the host
country decreases in emissions.

Next, the allowable donor country emission increases show a diminishing return to scale. As host country emissions are reduced, the allowable
increases in donor country emissions tend to become smaller. This effect is
echoed in diminishing sulphur exchange rates: the more abatement takes
place in the host country, the lower the exchange rate for subsequent abatement. The only case where this effect is absent is where no increase in grid
level deposition is allowed. But in this case, the allowable increase in donor
country emissions is very low – the associated exchange rate is only 0.02.

There are a number of consequences of these effects. The motivation for
joint implementation, it should be remembered, is to reduce sulphur emissions in the most cost-effective manner. For example, a donor country
might invest preferentially in emission reduction in a host country because
it is cheaper to reduce emissions abroad than to do so at home. But the reality of less than unitary (and falling) exchange rates means that emission
abatement abroad is considerably less attractive than might be expected
from simple cost comparisons (and becomes less attractive as time goes
on). Effectively, an exchange rate of (say) 0.5 means that the marginal cost
of abating emissions abroad must be less than half the marginal cost of
abating emissions at home, before the donor country will be tempted to
invest in emissions reduction preferentially in a host country.

Comparison with greenhouse gas abatement
What lessons can be learned from these considerations for the case of joint
implementation under the FCCC? Clearly there are quite considerable differences between the two situations (Table 5.1). In the first place, the problem of sulphur dioxide abatement retains essentially spatial aspects which
directly inform the relationships between donor and host in joint implementation arrangements. In the case of greenhouse gas abatement, the
assumption of uniform mixing of pollutants in the atmosphere means that
the spatial configuration of sources can be considered irrelevant to the
level of abatement carried out. Although the concept of exchange rates

Table 5.1: Differences with respect to issues associated with the JI of sulphur emissions and the JI of greenhouse gas emissions

Issue	Sulphur emissions	Greenhouse gas emissions
Status of science	Acidification science mature, although many uncertainties remain	Climate change science developing and major uncertainties remain
Environmental pathway	Usually treated as a non-uniformly mixed pollutant, i.e., spatial considerations important	Usually treated as uniformly mixed pollutants
Environmental impacts	Direct effects on human health, buildings and eco-systems and indirect effects through acidification	Predominately indirect effects through climate change processes
Mitigation	Source control only	Source control and sequestration
Legislation	Many decades of regulatory experience	Recent regulation only
Emissions trading	Experience of sulphur emissions trading in the United States; some experience of 'bubbles' in Europe	No large-scale carbon emission trading systems implemented
Importance of JI	JI is a relatively minor policy instrument within the contextof the CLRTAP and the Oslo Protocol at present	JI could become a significant policy instrument within the context of the FCCC and its future Protocols

Source: Authors.

(less than one) has been mooted for joint implementation under the FCCC, therefore, these exchange rates are not likely to evince the same spatial complexity witnessed under the Oslo Protocol.

Another crucial difference in the two institutional frameworks is the existence (under the Oslo Protocol) of national emission ceilings – and their current absence under the FCCC. National sulphur emission ceilings allow for the 'closure' of a joint implementation system. Trading takes place only between Parties to the Protocol, each of which has an initially designated emission ceiling. Thus increases (above target levels) in one part of the region, as a result of activities implemented jointly, take place only in the context of overall constraints on deposition levels at the grid-square level in the region as a whole. The current absence of emission ceilings under the FCCC means joint implementation arrangements cannot as yet be defined in terms of environmental targets. This makes the design of appropriate arrangements problematical, particularly with respect to ensuring that the environmental objectives of the Climate Convention are met.

In spite of these differences between the two situations, there are also some clear similarities (Table 5.2). In particular, both sets of pollutants are now the subject of major international reduction efforts, and both involve transboundary pollution issues. Perhaps more importantly, from the perspective of this paper, many of the sources of sulphur pollution are also sources of greenhouse gas emissions; and some at least of the technical avenues for sulphur abatement are also potentially avenues for greenhouse gas abatement.[10] This situation means that there are potentially double dividends in projects which simultaneously reduce both sulphur emissions and greenhouse gas emissions. At the moment, the institutional arrangements for activities implemented jointly are uncertain in both the CLRTAP and the FCCC. Consequently, it is currently unlikely that these potential synergies can be exploited.[11] However, there is clearly an incentive to remedy this situation, particularly when the incentives for investment in joint implementation arrangements for sulphur show diminishing returns to scale.

Conclusions

This chapter has outlined some prospective procedures for joint implementation under the Oslo Protocol to the CLRTAP. By comparison with

[10] This is not true of all potential solutions. For example, the use of fluidized bed combustion to reduce sulphur emissions can also increase carbon dioxide emissions.
[11] In fact, the separation of greenhouse gas-related environmental benefits from total environmental benefits implied by the GEF's existing concept of net incremental cost militates against such a synergy.

Table 5.2: Similarities in the issues associated with the JI of sulphur emissions and the JI of greenhouse gas emissions

Issue	Sulphur emissions	Greenhouse gas emissions
Multiple pollutants	SO_2, NO_x and NH_3 all contribute to acidification	CO_2, CH_4 and other greenhouse gases contribute to climate change
Transboundary pollutants	Efficient control strategies require internationally coordinated responses (although at the regional/continental level)	Efficient control strategies require internationally coordinated responses (at the global level)
UN Conventions	1979 UNECE Convention on Long-Range Transboundary Air Pollution	1992 Framework Convention on Climate Change
Economic activities and sectors	Combustion of fossil fuels dominant emission source and power/heat generation very significant	Combustion of fossil fuels dominant emission source and power/heat generation very significant
Clean technology and energy efficiency	Cleaner technologies and energy efficiency can reduce sulphur emissions	Cleaner technologies and energy efficiency can reduce carbon emissions
Economies in transition	Sulphur emissions have been reduced in European countries with economies in transition as a result of structural changes in economic/energy system	Carbon emissions have been reduced in European countries with economies in transition as a result of structural changes in economic/energy system
North European countries	Few energy supply-side sulphur abatement options remaining	Upward pressures on carbon emissions upon the energy supply sector
Application of economic instruments	Experience of emission taxes in Europe	Experience of emission taxes in Europe
Bilateral agreements in Europe	Experience of bilateral assistance to help reduce sulphur emission in countries with economies in transition	Experience of bilateral assistance to help reduce carbon emission in countries with economies in transition
Rules for JI	Rules and conditions for JI of sulphur emission reductions still being developed as part of the CLRTAP	Rules and conditions for JI of greenhouse gas emissions still being developed as part of the FCCC

Source: Authors.

the FCCC, the existence of clear environmental targets under the Protocol means that it is possible to define joint implementation arrangements without jeopardizing environmental objectives for reducing sulphur deposition exceedance levels. However, the spatial characteristics of acidic pollution introduce a level of complexity not encountered in the FCCC. This complexity militates against establishing simple exchange-rate arrangements between donor and host countries. Ultimately, an appropriately designed system would be likely to offer diminishing returns to scale, making international sulphur trading less and less attractive to potential donors. On the other hand, the existence of technical synergies between acid emission abatement and greenhouse gas abatement suggests that the institutional arrangements for the two kinds of joint implementation would be well advised to proceed at least in cognizance of, and perhaps in harmony with, each other.

Chapter 6

How does the sulphur market work?

Carlton Bartels

An SO_2 allowance essentially is almost a perfect commodity in the way it has been designed. It is perfectly fungible. If I have a client in the southern part of the United States who wants to buy an SO_2 allowance he can buy it from the utility in Portland, Oregon; he can buy it from the utility in Portland, Maine; he does not even have to know the utility. All the allowances live in an electronic database in the EPA so there is no movement or cost involved.

We have reduced the transaction cost so that someone in search of allowances need only develop their internal costs and understand their internal values, and then with one or two phone calls can price the entire market in the United States and call us up and say 'I need 10,000 tonnes; what does it cost?', and I will say 'Today it will cost you $91.50'. When we finally agree I can fax a confirmation to the buyer and the seller, they will send them back, the money will be wired over to me the next day, the allowances may already be held 'in street name' (i.e. by us as a broker). I believe at present I am holding about 350,000 allowances for about a dozen different parties, and we can clear a transaction in an afternoon – and that is remarkable.

I have a staff of seven people; my assistant and I deal with a volume of transactions which is two orders of magnitude greater than is dealt with by the other four people who are working in the ERC (the emission reduction credit) market. We can do it much faster and much more cheaply.

The evolution of the market was very interesting. First it was sort of foisted upon the utilities, who were willing participants when they saw what else was coming down. After an initial period of denial they have grown to develop the tools and embrace the market. When the programme came into place there were consultants running around saying 'I can help you, I will explain this programme', that gave way to investment banker-

type brokering where people would come and make very big and complicated transactions with large legal documents. I came in, in about the third
wave, saying 'There is no reason for this to be a big and complicated transaction with formal closings. It is just a purchase and a sale,' and I reduced
the whole thing to the point now where brokerage fees are about one per
cent of the value of an SO_2 allowance, and for large transactions they are
less than that.

Then the traders came in to add further liquidity to the market, because
now that you have a somewhat liquid market you can move these things
around. Transaction costs have collapsed; innovation is up; there are more
parties thinking of more creative ways of using allowances than ever
before, and in-house skills have grown.

When I used to make my calls three years ago I would be referred to the
Environmental Engineering department, which is the equivalent of a dead
end. These are the people who have been trained to minimize risk, to be in
compliance at all costs. Especially in the pre-competitive electric utility
industry, their main thing was to make sure they would not be found out of
compliance, especially since they were using their ratepayers' money.

These people would say 'My projected emissions are 10,000 tonnes but
I had better have 20,000 in the bank just in case.' Eventually this was transferred over to the fuel buyers and then the financial people, who say, 'You
know, I have a $100 million asset here and I will manage it.'

The users have expanded beyond the utilities. Now the fuel companies
are very active in this; some people find fault with this, the fact that fuel
companies may be buying allowances to bundle with higher-sulphur fuels
to make them more marketable. They miss the point that once the cap was
drawn, the line in the sand was drawn, and all the environmental benefits
of this programme have been won. Now we are dealing with economic
efficiency and moving the allowances to the highest value use, and if that
value use is keeping a mine open and maintaining employment in a certain
region, that is exactly where it should be. Those increases will be offset by
building a scrubber in the desert, or another plant.

Every time I hear about the magnificent savings that have occurred it
reminds me that when I was a child coming home from school I missed the
school bus, the public transport; I saw it go down the street and I ran and

just as I caught up it pulled away. As I pursued it I reached home before I ever got on the bus, and I was all hot and sweaty. My father said 'What happened to you?' and I said 'I ran home from school. I missed the bus, but I saved $1.50,' and he said, 'Chase a taxi next time, you'll save $5.' So the price savings are very subjective when we get down to it.

There is no doubt in my mind that when you take the decision-making out of the central core of the bureaucracy and put it out into the market-place, you will find cheap reductions in these programmes that were not expected. I remember an early conversation with one of my clients, saying 'What is your cost of reducing sulphur dioxide?' and he would never tell me. I was trying to make the point that you can sell these things. Finally when we got to know each other and became comfortable he said, 'Actually we save about $5 a tonne.' I said, 'What do you mean, you save?' and he said, 'Well, there was always this lower-sulphur fuel we could buy, but under the old regime we were not going to buy it even though it was less expensive because we were afraid that if the price reverted we could never go back to the higher-sulphur fuel, because under the old programme it would be a one-way ratchet. Therefore we were pol-luting at a loss.'

This is appalling, but it makes perfect sense. Why should someone go and do something for which they may only be penalized? The motivation issue in trading is incredibly important and one of the problems that I have with JI is that I have yet to see any motivation for a business to engage in it; that is the way it is set up. The benefactor is the government, but what do I get out of it? This is why I have not staffed up yet. We have to make sure that the players have the opportunities.

My own view on translating SO_2 to CO_2 is that you have to be ready for the programme, and you have to be a willing participant. I have been advo-cating that the United States could think about doing this as a way to meet its own internal domestic needs. If that were to occur and JI or something else were a mechanism to allow other countries to participate and bring credits into the United States, there would be a cash value established and there would be a baseline from which other countries might start saying, 'This makes sense. These people are paying money for this stuff'; and when they are paying money you will find all sorts of people willing to sell.

We have been looking for carbon credits, we have a buyer for CO_2 offsets right now for very unusual and specific circumstances, but they are willing to buy millions of tonnes and they are willing to pay money for them. We have been able to present them with projects domestically at about 25 cents a tonne of CO_2 equivalent; this is in millions of tonnes, things that are available and are real. It just shows that there are a lot of opportunities out there, and we are getting there.

While a programme is being implemented you can find a million reasons to make it complicated which five years of experience will show were unnecessary and made it much more difficult; once people have experience with the programmes they will say, 'Why did we do all that?' Phase I and Phase II of the SO_2 programme are a marvellous example of this lesson.

We have created more problems than we solved. Markets can be made rather simply and elegantly. If you make a programme with a firm commitment and people have a motivation to trade, the markets will take care of the efficiency. All you have to do is to make sure there are enforcement and monitoring provisions and a reason.

Once you start and become involved and engaged, you can say, 'This really works out very nicely, and we like this flexibility.'

Chapter 7

The evolution of marketable permits: the US experience with sulphur dioxide allowance trading

Brian J. McLean

Introduction

In the pursuit of environmental protection, government policies and programmes should strive to be as effective and efficient as possible. The Sulphur Dioxide (SO_2) Allowance Program, created under Title IV of the Clean Air Act Amendments of 1990,[1] represents a radical departure from the traditional command-and-control approach to environmental policy and regulation and from previous emissions trading efforts. It contains the first statutorily mandated, national, market-based approach to environmental management. Its implementation, therefore, provides a unique opportunity to assess the effectiveness of such an approach. Although initial emissions reductions were not required until 1995, with full implementation not until the year 2010, the first six years since enactment provide a wealth of experiences from which some early lessons can be learned about designing and implementing market-based approaches for protecting the environment.

This chapter reviews the goals of the SO_2 Allowance Program and the results after its first full year of operation, discusses the context for the establishment of the programme and its critical design features, and then identifies some lessons learned from implementing this large-scale emissions trading programme.

Goals and results

Although SO_2 emissions in the United States had declined from their peak of 33 million tons in 1973 to 25.9 million tons in 1980, the 1990 Clean Air Act set a goal of further reducing SO_2 by 10 million tons below the 1980

---

[1] Clean Air Act Amendments of 1990 (Public Law 101-549), 15 November 1990.

level to protect public health and the environment. It was expected that emissions other than those coming from the electric power sector would decline by 1.5 million tons (primarily from replacement of older, higher emitting facilities with new, lower emitting facilities) and that electric utility emissions would need to be reduced by 8.5 million tons. The reduction in utility emissions was to be accomplished as cost-effectively as possible through the application of a system of tradeable 'allowances', where one allowance would be a limited authorization to emit one ton of SO_2. Emissions reductions were to begin in 1995 with full compliance expected by the year 2010.

For the first phase of the programme (1995–9) only 263 boilers out of the more than 2,000 fossil-fuel-fired boilers and combustion turbines used to generate electricity in the United States were subject to emissions limitations. However, because of provisions in the law which allowed Phase II combustion units to participate voluntarily in Phase I, 182 additional units, or a total of 445 units, were affected in 1995, the first year of the programme.[2]

In 1995, these 445 utility units reduced their SO_2 emissions to 5.3 million tons from their 1980 level of 10.9 million tons, with most of that reduction occurring in 1995. Since 8.7 million allowances were issued to these sources for 1995, they emitted 3.4 million tons (or 39 per cent) below their allowable emission level for that year (Figure 7.1). The 5.6 million ton reduction from Phase I utility units, coupled with a reduction of more than 2 million tons from non-utility sources, led to a total reduction in SO_2 emissions of 7.9 million tons from the 1980 level (Figure 7.2).[3]

The expected costs of the programme have declined since it was debated in Congress. Early estimates of the cost ranged from $180 to $981 per ton of SO_2 removed during Phase I and from $374 to $981 per ton during Phase II.[4] For the past year the cost of buying a ton of SO_2 reduction (for

---

[2] *1995 Compliance Results: Acid Rain Program*, EPA 430-R-96-012, US Environmental Protection Agency, Washington, DC, July 1996.
[3] *National Air Quality and Emissions Trends Report*, 1995, EPA 454-96-008, US Environmental Protection Agency, Washington, DC, December 1996.
[4] Hahn, Robert W. and Carol A. May, 'The Behavior of the Allowance Market: Theory and Evidence', *The Electricity Journal*, March 1994, pp. 28–37.

Figure 7.1: SO₂ emissions, 445 Phase I affected utility units

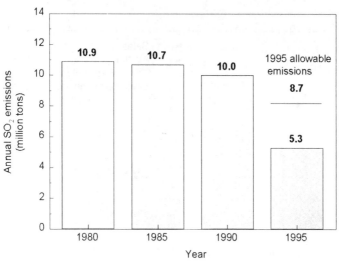

Source: Authors.

Figure 7.2: National SO₂ emissions, all sources

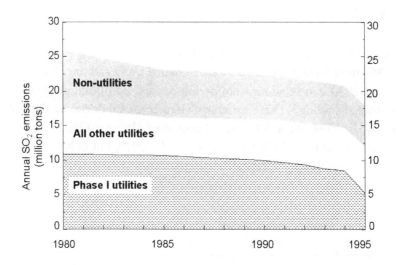

Source: Authors.

Figure 7.3: SO$_2$ allowance prices

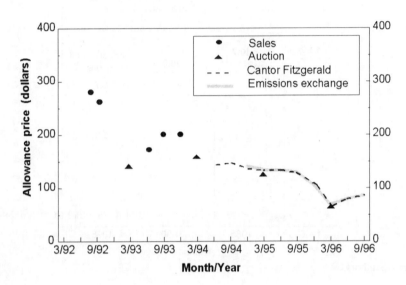

Source: Authors.

either Phase I or Phase II compliance) has been less than $100 (Figure 7.3). In 1990, total annualized costs were estimated by EPA to be $5 billion by the year 2010 if no trading were permitted, and $4 billion per year with unrestricted trading.[5] These estimates were considered low by the utility industry. In 1994, the US General Accounting Office estimated the costs without trading to be $4.9 billion per year by 2010, but only $2.0 billion per year if full trading occurred.[6]

---

[5] *Comparison of the Economic Impacts of the Acid Rain Provisions of the Senate Bill (S1630) and the House Bill (S.1630)* (Draft), prepared by ICF Resources Inc., US Environmental Protection Agency, Washington, DC, July 1990.

[6] *Air Pollution: Allowance Trading Offers an Opportunity to Reduce Emissions at Less Cost*, GAO/RCED-95-30, US General Accounting Office, Washington, DC, December 1994.

There appear to be several explanations for the lower-than-expected compliance costs. First, the allowance system facilitates competition across all emission reduction options. Flue gas desulphurization, a major control technology option, costs 40 per cent less than before the 1990 Clean Air Act; removal efficiencies have also increased from 90 per cent to 95 per cent or more. Productivity at both low- and high-sulphur coal mines has continued to improve at rates exceeding 6 per cent per year, and rail transport tariffs, which had declined somewhat before the Clean Air Act, dropped by 40 per cent after the Act was passed.[7]

Second, market systems with their competition and flexibility provide incentives for innovation. Companies are experimenting with fuels for which their boilers were not designed and blending fuels to minimize SO_2 emissions. Also, coal suppliers are 'bundling' allowances with coal sales to increase their attractiveness. Market instruments such as options and swaps are being employed to reduce risk.

Third, in addition to trading, which allows the units with the lowest compliance costs to bear the burden of control, the banking provision has provided considerable flexibility in timing emissions reductions, minimizing operational disruptions and allowing capital expenditures to be delayed. Finally, a market system can reveal the true costs of compliance, informing the market participants, who, in turn, can make more cost-effective compliance choices.

The low cost being revealed only a few years after the high estimates were made raises other important questions. Why were the estimates of almost all economists and analysts so high? How can we do a better job of predicting costs for future programmes? And what weight should estimates by the regulated industry be given in debating and designing future programmes?

Besides being less expensive for sources to meet their emissions reduction obligations, the programme can be less expensive for the government to administer. In the first five years of the SO_2 Allowance Program (which included programme development and the first year of operation), govern-

---

[7] US Environmental Protection Agency, *Analyzing Electric Power Under the Clean Air Act Amendments*, July 1996.

ment expenditures amounted to less than $60 million out of a total esti-
mated government expenditure for air pollution control of $3.5 billion.
Approximately 15,000 people work on air pollution control for federal,
state and local governments. Because the Allowance Program is heavily
automated, less than 150 workyears per year are expected to be needed to
operate it. It is likely that about three-quarters of that workforce would
focus on auditing the performance of emissions monitors and assuring the
quality of data reports. The remainder would handle all other functions,
including permitting, allowance transfers, allowance auctions, data system
operations and enhancements, end-of-year allowance reconciliation, programme evaluation, and general administration.

Context and design

To understand why the programme has been successful, it is necessary to
understand the context in which it was developed and critical design features embodied in it.

Context

The seeds of successful implementation are often laid long before actual
implementation. In this case, considerable scientific investigation, economic and implementation analyses and debate preceded enactment of
Title IV. As a consequence of this extensive and often innovative research,
it was determined by 1989 that anthropogenic sulphur dioxide emissions
were the primary cause of acidified lakes and streams and visibility degra-
dation in the eastern United States, and a contributor to materials damage
and health effects.

In addition to more than ten years of scientific investigation, there had
been a comparable period of policy debate over what should be done about
SO_2 emissions. In fact, by 1989 more than 70 bills had been introduced in
Congress specifically addressing the acid rain issue. Accompanying the
policy debate was the development and improvement of analytical tools to
assess the economic impacts of the various proposals. Sophisticated linear
programming models, such as the Coal and Electric Utility Model

(CEUM), were used by industry, government and environmental organizations to test and develop a multitude of policy options and alternatives.

All the economic analyses indicated that any significant reduction of SO_2 would cost billions of dollars and affect thousands of high-sulphur coal mining jobs. In 1986, Congress and the administration agreed that a public/private partnership to commercially demonstrate technologies for burning coal more cleanly was appropriate, and the Clean Coal Technology Program was begun.[8] Since its inception, $7 billion in public and private funds have supported 45 projects.[9]

In 1984, EPA launched an unprecedented effort to study the potential implementation issues posed by any attempt to significantly reduce SO_2 emissions under an acid rain control programme. By 1988, 36 state and local agencies had participated with EPA in the State Acid Rain (STAR) programme which identified key components for a successful implementation effort should legislation be enacted.[10]

In 1987, EPA also invited state public utility commissioners, through the National Association of Regulatory Utility Commissioners (NARUC), to assess issues and approaches to implementing any acid rain control effort. Early in 1989, EPA brought together directors of state air pollution control agencies and public utility commissioners to explore opportunities for coordination should acid rain legislation become reality.[11]

For the ten years leading up to legislative action, the acid rain issue benefited from one of the most intensive and expensive ecosystem research efforts, sophisticated and extensive economic analyses, unprecedented pre-legislative federal/state implementation analyses, and substantial financial support to demonstrate cleaner and cheaper control technologies.

---

[8] *Joint Report of the Special Envoys on Acid Rain*, Drew Lewis (US), William Davis (Canada), January 1986.
[9] *Comprehensive Report to Congress; Clean Coal Technology Program: Completing the Mission*, DOE/FE-0309P, US Department of Energy, Washington, DC, May 1994.
[10] *State Acid Rain Program; Final Report*, EPA 400/1-89-001, US Environmental Protection Agency, Washington, DC, March 1989.
[11] *Report on Workshop Proceedings; Acid Rain Control: How States Respond, A Workshop for Public Utility Commissions and State Air Agencies*, 30–31 January 1989, prepared by RCG/Hagler-Bailly, US Environmental Protection Agency, Washington, DC, 23 May 1989.

Nevertheless, in 1989 the cost of control was still considered politically unacceptable. To break the log-jam, a market-based approach was introduced which held the possibility of substantially reducing those costs. This raised new questions:

- Could it achieve the promised cost savings?
- What would be the environmental consequences of letting market forces decide where emissions reductions would take place?

Fortunately, in the case of acid rain, the geographic, economic and environmental circumstances were conducive to using a market-based implementation approach, specifically an emissions trading programme.

- Economic and environmental analyses indicated that under an emission trading programme emissions reductions were likely to occur where they would be most beneficial to the environment.
- Environmental analyses indicated that annual emissions reductions would be sufficient to provide benefits; hourly or daily limits were not critical.
- A variety of cost-effective technological, fuel switching, and energy efficiency options existed to reduce national SO_2 emissions by 50 per cent or more.
- The ubiquitous nature of the electric utility industry allowed compliance costs to be spread broadly.
- Electric utilities, which contributed about 70 per cent of the SO_2 in the United States, reported considerable environmental and economic data, facilitating analyses and equitable implementation.
- The technology existed to accurately measure emissions, providing the accountability and credibility essential for a market approach.

Programme design – starting from scratch
Unlike previous emissions trading programmes, which simply added flexibility to the underlying command-and-control infrastructure, the concept with Title IV was to build an entirely new programme that would be com-

pletely separate from other requirements. While health-based standards for sources would remain in place, they would not affect the goals of the new programme or control any of the aspects of the emissions trading. In other words, the allowance programme would be built on the experience of previous command-and-control programmes and emissions trading programmes, but not on their infrastructure.

In most situations, therefore, emissions trading required the conversion of these requirements into a tradeable commodity, such as tons of a pollutant emitted. This required all parties (emissions credit generator, consumer and governmental authority) to agree on the present (baseline) and future utilization rates for generating and consuming sources, to agree on the time over which the trade would be valid, often to agree that the reductions for which credit was being given would not have happened but for this trade, and to agree on how emissions at both sources would be quantified.

This process could be resource-intensive and time-consuming, with changes to the agreement often requiring renegotiation. Despite the considerable efforts to ensure that transactions would not worsen the environment, most air regulators and environmentalists remained suspicious of emissions trading.

With over ten years of emissions trading experience to draw from, the SO_2 allowance programme created by EPA and enacted through the 1990 Clean Air Act attempted to maximize the economic advantages of emissions trading by minimizing transaction costs and to build environmental credibility by reducing and capping total emissions and holding sources accountable for every ton of emissions.

Key features

Goals. The difference between the SO_2 allowance programme and previous air pollution control efforts begins with the definition of the goals and source obligations. The programme objective was established as a maximum annual emissions level (or permanent 'cap') of 8.95 million tons. To achieve this, only 8.95 million allowances (the legal authorization to emit a ton of SO_2) would be issued each year, and a source would have to hold an allowance for each ton it emitted. This is the first air programme to limit total emissions, and the first to rely exclusively on emissions (output) for determining compliance, as opposed to relying on a specific technology, emissions rates (mass per unit of heat input), concentrations, or percentage reduction, all of which do not limit total emissions. This meant that government and industry had the same goal: to limit emissions. It also allowed plants to get full credit for emissions reduced through improved efficiency or lower utilization.

Trading. Instead of the credits being calculated as reductions (either projected reductions, or after they occur) as in previous trading programmes, the trading unit is defined as the allowance, i.e., one ton of allowable emission. Allowances are issued to sources on the basis of a series of formulae reflecting recent historical utilization and desired emissions rates, generally issued in perpetuity, and before the trading of allowances or emission reductions begins. In this way, the cost of the trading infrastructure is moved upfront. To maintain the emissions cap, new sources must acquire allowances from existing allowance holders or through the government auctions (which sell 2.8 per cent of allocated allowances). To promote the use of market intermediaries and create an incentive for early compliance with the emission reductions, allowances may be traded to any party and may be banked for use in a subsequent year, but they may not be brought forward for use in an earlier year. EPA maintains the central registry of allowances (containing 31 years' worth of allowances for each account) and provides a quick turnaround of transactions (usually within a day).

Emissions measurement. Along with the improvements in the allowance market, a common measurement metric was created through continuous emissions monitoring systems (CEMS), with quarterly reporting of hourly emissions to EPA. After the end of each year the total number of tons of SO_2 emitted by each boiler is then deducted from allowances contained in each electric utility unit's account, with any excess allowances rolled into the following year's account.

Permitting. Because all emissions are measured and no detailed compliance schedules are needed, permit applications are much more streamlined than traditional operating permits. For the allowance programme, the applicant must simply provide the name of the plant, commit to measure emissions, and commit to hold sufficient allowances to cover annual emissions. These changes dramatically reduce the cost of permitting for both the source and the government and increase the likelihood of sources seeing the allowances as a full compliance alternative.

Automatic non-compliance penalties. Coupled with the requirements to measure emissions accurately is the mandated, immediate penalty for non-compliance. If annual SO_2 emissions exceed the number of allowances held at the end of the year, statutory penalties of $2000 per ton exceeded (indexed to inflation) and an offset of one allowance per excess ton are assessed automatically.

Default limits. If EPA failed to put implementing regulations in place, emission limitations stated in the law would apply to every source (with no reallocation of emissions limitations possible). Coupled with the automatic non-compliance penalties, this provision encouraged the industry to support the timely promulgation of regulations to avoid the more costly statutory fallback. The requirement to measure all emissions and the automatic non-compliance penalties and default limits were not part of the existing air pollution control infrastructure. They are not only critical to the credibility and efficiency of the emissions trading aspect of the allowance programme, but they enhance its environmental integrity.

Other protections. Because emissions trading allows for flexible emissions patterns, a common concern with programmes of this type is that a 'hot spot' of emissions will occur through trading, impairing health or welfare in the local area, even as the larger goal of regional emission reductions is attained. All utilities regulated under this programme are also required to comply with all other requirements of the Clean Air Act, in particular the requirement to meet the health-based National Ambient Air Quality Standards, New Source Performance Standards, and Prevention of Significant Deterioration provisions. These requirements are independent of the trading programme and cannot be circumvented through the purchase of allowances.

Programme implementation

Status of implementation

As of March 1993, EPA had issued all the rules necessary to implement the SO_2 Allowance Program:

- permitting of the utility plants;
- allocations of the SO_2 allowances for all utility units;
- procedures governing transfers of allowances;
- continuous emissions monitoring requirements;
- the rules covering non-compliance penalties and administrative appeals.[12]

In April 1995, EPA issued rules allowing industrial boilers and small utility units to enter the allowance trading system voluntarily.[13] Litigation on aspects of permit rules, monitoring rules and allowance allocation were resolved relatively quickly . As rules were completed, the emphasis shifted to programme implementation. The SO_2 allowance programme includes over 2,000 existing boilers and turbines serving electric generators throughout the United States (except for Alaska and Hawaii), ranging in size from 25 megawatts (MWe) to 1300 Mwe, and will include all new combustion devices serving commercial electric generation. Approximately 1,000 of the units burn coal of various types and in a variety of boiler configurations. The remainder use oil and/or natural gas.

The utility industry met all permitting deadlines for the 445 Phase I units. EPA reviewed the plans and issued all permits by the end of 1994. Utilities installed and tested continuous emissions monitoring systems at affected units (both Phase I and Phase II), and EPA certified them by the end of 1996. The accuracy of the instruments has been considerably better than

---

[12] US Environmental Protection Agency, *Acid Rain Program: General Provisions and Permits, Allowance System, Continuous Emissions Monitoring, Excess Emissions and Administrative Appeals; Final Rule*, 58 FR 3590, 11 January, 1993. US Environmental Protection Agency, *Acid Rain Allowance Allocations and Reserves; Final Rule*, 58 FR 15634, 23 March 1993.

[13] US Environmental Protection Agency, *Opting Into the Acid Rain Program*, 60 FR 17100, 4 April 1995.

Figure 7.4: Allowances transferred

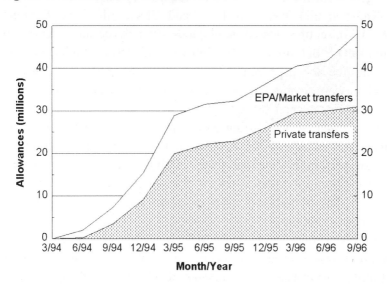

Source: Authors.

Figure 7.5: Private allowance transactions recorded, 3/94 – 9/96

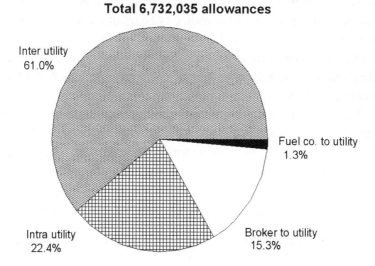

Source: Authors.

the stringent performance specifications required by the rules.[14]

Allowance trading began in 1992, and EPA's Allowance Tracking System (ATS), set up to record those trades officially, became operational in March 1994.[15] Because of the flexibility provided by the trading system, sources are free to engage in a variety of allowance transactions, such as cash and forward contracting and options trading, without interference from or needing the approval of EPA.

Allowance trading is occurring on two levels, both of which result in significant cost savings to the affected industry. First, most large utility systems are performing 'internal' trades of allowances among the boilers within their systems in order to achieve the lowest cost of compliance for their system. The second level is the external trading among utilities and various financial entities.

As of 28 November, 1996, over 1,700 private transfers involving over 30 million allowances had been recorded in ATS (Figure 7.4).[16] Most of these transfers have been reassignments of allowances among multiple owners or movement in and out of general accounts. We have categorized 6.7 million of these transferred allowances as involving private trades (Figure 7.5). Transfers are processed through presentation of a simple form to EPA detailing the allowances to be transferred. Ninety per cent of the transfers have been recorded within 24 hours of receipt. Since submission of trades for recording is voluntary and will not occur until the parties want the transfer to be 'official' for compliance purposes, market watchers speculate that significantly more allowances are being 'traded' than are being recorded in ATS.

The statute provided for annual allowance auctions to begin in 1993. Through the Chicago Board of Trade, EPA has conducted four auctions at which a total of 775,000 allowances have been sold. The auctions have been useful in getting the market started and in helping to reveal market

---

[14] US Environmental Protection Agency, *Acid Rain Program: Update No. 2: Partnership for Cleaner Air*, EPA 430-N-95-012, July 1995.
[15] US Environmental Protection Agency, *Acid Rain Program: Allowance Tracking System*, 59 FR 10131, 3 March 1994.
[16] US Environmental Protection Agency, *Transaction Summary of the Allowance Tracking System*, 28 November 1996.

prices. As private trading has increased, though, they have become a smaller part of the allowance market, and private services are providing more frequent information on allowance prices.

In 1994, Phase I units began reporting hourly emissions of SO_2, NO_x and CO_2 to EPA's Emissions Tracking System (ETS), which is one of the largest data systems used by EPA to manage an environmental programme. Phase II units began doing so in April 1995. EPA periodically reports the plant-by-plant emissions of SO_2, NO_x and CO_2 for these units as measured by their continuous emissions monitoring systems.[17]

Since 1993, through the special Conservation and Renewable Energy Reserve of allowances created by Title IV, EPA has awarded 18,503 bonus allowances to 27 different electric utilities for demand-side energy efficiency measures and for renewable energy projects. Under other provisions of Title IV, EPA has issued 3.5 million allowances to Phase I utilities that are employing flue gas desulphurization systems, 314,248 allowances for Early Reductions Credits, and 66,969 allowances to refineries for desulphurizing diesel fuel. In 1996, EPA issued the first two permits to industrial facilities that wished to voluntarily join the allowance trading system under the opt-in programme.

Finally, in July 1996, EPA completed the first annual reconciliation process where each unit's annual emissions are compared to its allowance holdings to determine compliance. All 445 units participating in the programme in 1995 were in compliance, and no penalties were assessed or offsets required.

Partnerships, dialogue, and new ways of doing business

To implement the SO_2 Allowance Program successfully, EPA has pursued partnerships, dialogue, and new ways of doing business with industry and the environmental community. Immediately following enactment, with the programme's goals, structure, and responsibilities clearly established, EPA initiated an intensive dialogue with the affected stakeholders through the

---

[17] US Environmental Protection Agency, *Acid Rain Program: Emissions Scorecard 1994*, EPA 430/R-95-012, December 1995.

Acid Rain Advisory Committee (ARAC). The ARAC was composed of 44 individuals representing a variety of different interests, including large and small utilities, coal and gas interests, state air agencies and public utility commissions, environmental organizations and academia.

For six months, the members of the Advisory Committee became actively involved in devising solutions to problems and offering critiques of various regulatory options for implementing Title IV. With stakeholders well represented and the meetings well attended by the public, potential problems were identified early on. Perhaps most importantly, ARAC was helpful beyond the rule-making process; it helped facilitate implementation because it produced a cadre of knowledgeable individuals who were committed to making the programme work.

After publication of the rules, dialogue continued. In the area of permitting, pre-application meetings served as a primary vehicle for EPA–industry dialogue. These meetings, held across the country, provided opportunities for utilities to ask questions and received written responses from EPA headquarters and regional personnel.

In both monitoring and data system development activities, EPA hosted numerous training sessions around the country and attended industry-sponsored meetings. EPA not only provided policy guidance in these areas, but developed and distributed software to assist industry in developing their emissions reporting systems.

To set up an allowance market not only required maintaining communications with the utility industry, but it also meant establishing ongoing relationships with the financial community and public utility commissions. Outreach activities have included participation in EPA and industry-sponsored training and conferences, discussing rate-making issues with rate regulators, holding the annual allowance auctions through the Chicago Board of Trade, and disseminating information on the auction process, energy conservation, renewable energy, and the allowance and emissions tracking systems.

Programme expenditures

During the first five fiscal years since enactment of the 1990 Clean Air Act Amendments (FY91–95), EPA expended $1.09 billion to implement the act. Of this amount, $44 million was spent by EPA to implement the acid rain programme, approximately $38 million of which was spent on the SO_2 allowance programme.[18] EPA also awarded $833 million over these five years to state and local governments to implement the Clean Air Act, of which $18.9 million was allocated for acid rain programme implementation. State and local governments also provided significant support to air pollution control efforts.

Programme development and support ($19.9 million) was the largest area of investment in the first five years and included development of rules, guidance and procedures; response to litigation; state permit programme approval; and all ongoing support activities that did not involve actual operation of the programme. Data system development ($9.9 million) included all contract and personnel support for designing and constructing the two major data systems – the Emissions Tracking System and the Allowance Tracking System. Programme operations ($10.8 million) included those activities involved in actual implementation.

The 'Other' category ($3.5 million) included performing certain statutorily mandated studies, such as the Acid Deposition Standard Feasibility Study and cost-benefit analyses to support requirements under sections 812 and 901 of the Act, and included participation in the National Acid Precipitation Assessment Program. The 'Other' category also included general outreach and communications activities, programme evaluation and implementation of the US–Canada Air Quality Agreement.[19]

Expenditures shifted over the five-year period. In the first two years, almost three-quarters of expenditures were for programme development. By FY95 priorities had shifted significantly to where only 24 per cent was devoted to programme development and support, 27 per cent to data system

---

[18] Brian J. McLean, *The Sulphur Dioxide (SO₂) Allowance Trading Program: The First Five Years*, US Environmental Protection Agency, Washington, DC, 19 January 1996.

[19] Agreement between the Government of the United States of America and the Government of Canada on Air Quality, signed by the United States and Canada, March 1991.

development, and 43 per cent to programme operations. If funding remains stable, these relative proportions should hold for the next two fiscal years as key rule-making and data system improvements are completed.

Lessons learned

At this point, after around two years of operating experience, several observations can be made about the SO_2 allowance programme.

(1) Traditional operating permits can be greatly simplified. With rigorous emission monitoring and the flexibility of an allowance system there is no need for setting source-specific emission limits, specifying control technology, or requiring detailed compliance schedules.

(2) Transaction costs can be very low because government involvement in an allowance transaction simply involves recording, not case-by-case review or approval, and the source has numerous venues in which to transact allowances. There are hundreds of allowance holders, several brokers, and the annual no-fee government allowance auctions. Furthermore, the decision to trade is not a stand-alone one, but part of a company's overall compliance strategy.

(3) The ability to bank allowances can reduce costs and lead to significant early emissions reductions, but it can also extend the time for achieving the ultimate emission reduction target.

(4) Continuous emissions monitoring can be expensive, particularly for the installation of elevators and platforms for tall stacks and for frequent quality assurance, but accurate monitoring and timely reporting are also critical to the credibility of the entire trading programme, and their cost is modest when compared to the overall cost savings. The cost of monitoring SO_2, NO_x and CO_2 at over 2000 sources has been estimated at about \$200–300 million per year, compared to the cost savings for the SO_2 programme alone of \$2–3 billion per year.

(5) Phasing in the participation of sources can complicate administration and undermine achievement of emission reduction goals and has been perhaps the most serious flaw of the SO_2 Allowance Program. Two types of problems can occur: (a) participating sources can shift electri-

cal load to non-participating sources, and (b) voluntary participants can earn allowances that may be used by other participants in lieu of reducing emissions, while the non-volunteering sources increase their emissions and cause a net rise in emissions. Administrative mechanisms to compensate for these problems can be complex and are of limited effectiveness in ensuring the environmental integrity of the programme. The 'substitution' and 'reduced utilization' provisions employed in the SO_2 Allowance Program have been litigated and revised, and have become the most complicated administrative parts of the programme. Approximately 75 per cent of the cost of developing and implementing the permitting provisions of Title IV and at least one-third of the cost of developing and operating the allowance tracking system, or about $6.6 million, can be attributed to the complexity of Phase I. In retrospect, all affected sources should have been included from the outset in Phase I with emissions limitations tightened in Phase II to accomplish the goals of the programme.

(6) Capping total emissions coupled with allowance system compliance flexibility is compatible with electric power industry restructuring. The electricity industry is the largest source of SO_2 in the United States and one of the largest sources of NO_x, mercury deposition, fine particulates and CO_2. The industry is also currently undergoing a major restructuring from domination by regulated monopolies to competing nationally on price and availability. Traditional air pollution control programmes assume stable patterns of electricity production. Because they have no emissions caps, emissions could increase as well as shift geographically, if, as a result of restructuring, production shifts to higher emitting plants with lower pollution control costs. A cap and trade programme accommodates a dynamic market situation by ensuring that total emissions will not increase and by allowing costs of emissions control to follow shifts in production and emissions.

(7) Cost savings can exceed expectations. Since 1990, the projected cost of compliance with the full SO_2 emission reductions has declined from $4 billion per year to less than $2 billion per year, against an annualized cost of compliance without trading of $4.9 billion. Although some of the cost savings can be attributed to the unexpected lack of increase in

fuel prices, competitive markets do continuously seek more cost-effective solutions, leading to more rapid innovation and cost savings.
(8) Governmental administrative costs can be much lower than traditional programmes. By streamlining permitting, eliminating case-by-case review of trades, removing government participation in compliance decisions, and focusing instead on the measurement of emissions produced by affected sources, considerable public resources can be saved.

One governmental decision not eliminated by the allowance trading system is the need to allocate allowances. This usually requires some consideration of historical utilization and emissions information. For traditional programmes, historical (and sometimes future) utilization and emissions information is required as each source receives an emission limit or applies for approval of a trade. For an allowance trading programme this happens once for all participating sources at the beginning of the programme. The advantage of this is more equitable and consistent treatment of sources, and elimination of what have often been lengthy delays in the approval of trades.

For the SO_2 Allowance Program, most of the allocation decisions were made by the US administration and Congress in 1989 and 1990 as part of the legislative process. Initially a few allocation formulae were established to recognize significant differences among existing sources, e.g., stemming from different fuels and historical levels of control. But this blossomed into 29 formulae, as Congress recognized that special consideration could be given to special situations. Written by different people for different interests, the language was inconsistent and ambiguous, demanded numerous special data requirements, and served neither the environment nor market efficiency. As a result, the allocation process became unnecessarily costly and long, and provoked litigation. Approximately one-third of the cost of developing and supporting the allowance trading programme, or about $1.4 million, can be attributed to this factor. Allowance formulae and data requirements can and should be kept to a minimum, with requirements defined clearly and consistently.

Conclusions

From the SO_2 Allowance Program we have learned several lessons about designing and implementing effective and efficient environmental programmes.

(1) Actions prior to programme design are important. This is where our best scientific understanding is brought to bear on defining the problem and setting goals, where the economic, social, and environmental implications of alternative solutions are evaluated, and where the appropriateness of using a market-based approach is explored. It is also during this stage that the public's acceptance and support of the proposed solutions is assessed. Market-based instruments are tools to solve problems, but first the problem must be defined, goals set, and the need to take action accepted.

(2) The design of the programme is critical because it determines whether effective and efficient implementation is possible. The goals and responsibilities should be clearly stated, and there should be unequivocal consequences for not complying or for delaying implementation. From the beginning of the programme the emissions of all potentially affected sources should be accounted for and a maximum allowable emissions level (or cap) should be established (and sustained). Accurate measurement of emissions is the key to environmental accountability, market credibility, and operational flexibility. Allowance allocation is primarily a political process, not an environmental one. Overall, the design should be simple. This will ensure faster start-up, greater certainty for all stakeholders, and lower administrative cost.

(3) In implementing the programme, government should stay focused on achieving the goals in the legislation, resolving issues promptly, and improving operational efficiency. It should refrain from trying to participate in, control, or fine-tune the market, particularly since many changes, such as restructuring, may occur outside the regulator's purview. This focus should provide the certainty, efficiency and stability desired by all and necessary for optimal market performance. All those involved in implementation – government and industry – should

maintain a vigorous dialogue with the goal of continuous improvement in both the environmental effectiveness and the operational efficiency of the programme.

Even though it has been only six years since its enactment, the SO_2 Allowance Program is providing useful lessons for the effective and efficient protection of public health and the environment. It also serves as a benchmark against which other air pollution control and market-based programmes can be measured.

Chapter 8

Controlling carbon and sulphur international investment and trading

Raúl A. Estrada-Oyuela

As is well known, the negotiation of the Framework Convention on Climate Change was triggered by science, particularly by the IPCC report to the Second World Climate Conference in November 1990. While the scientific segment of that conference was developing in one building in Geneva, some of us were discussing the conditions of the negotiation in one of the rooms of the Palais des Nations.

From then on, the whole process has been highly dependent on the inputs received from the IPCC, as the major and most relevant source of scientific information and assessment. In March 1993, in New York, immediately after being elected Chairman of the Intergovernmental Negotiating Committee whose mandate was extended by the UN General Assembly in order to help the entry into force of the Convention, I sent a letter to Professor Bolin, Chairman of the IPCC, spelling out the foreseeable scientific needs of the Convention's process.

I can't say that my request was refused, but the answer was that the IPCC had not been created to deliver upon requests, that it had its own programme and method of work. We started a long process of dialogue; cooperation is going on today. We held a series of 'joint-Bureaux' meetings and the fact is that in the week of 8–14 December 1996, the 5th session of the Ad Hoc Group on the Berlin Mandate (AGBM) profited from the content of a document on technologies, policies and measures for mitigating climate change, custom-tailored by the IPCC for AGBM.

At the same time, it is a fact that scientific assessments in matters related to climate change were affected by political interests, at least affected in a way that impeded or hindered more concrete definitions. Some points in the summary for policy-makers of the Second Assessment Report were literally 'negotiated' in Rome in December 1995.

This is a very curious process, because the Ministerial Declaration adopted during the Second Conference of Parties (COP-2) in Geneva in July 1996, which is a political document, drawing from the IPCC's Second Assessment Report, reflects the scientific community's concerns in a very assertive way: the ministers say that without mitigation, the temperature could rise between 1° and 3.5° C and sea levels could rise between 15 and 95cm by the year 2100.

Of course that was the Ministerial Declaration, but it was not the declaration of all ministers attending the COP-2. Ministers representing oil-exporting countries dissociated themselves from the document which, for that reason, was not 'adopted' but 'noted' by the Conference.

The political reading of these reactions to the IPCC's Second Assessment Report is simple: interests play a very relevant role in decisions, and this is not going to be changed by invoking the precautionary principle. It is necessary to understand that AGBM negotiation draws from scientific and technical sources, but it is a political exercise.

We do not have scientific definitions peacefully accepted by all parties, nor do we have clear-cut technical solutions that governments can simply verify and adjust over time to their own countries. We have only 'open-ended' assessments, subject to verification and further analysis, and catalogues of possible policies and measures and reductions and limitations that cannot be applied globally without nuances.

With superb cooperation from the Secretariat of the Climate Change Convention, I prepared a synthesis of proposals by the Parties, which I had undertaken to produce during the 4th AGBM session in July 1996. That synthesis identified the issues and the ways and means suggested to solve them. I did my best to understand the substance of different proposals, to organize them in a comparative way and, whenever possible, in a compatible way.

The Berlin Mandate is addressed to all Parties to the Convention, but in two different ways. Because the commitments under Art. 4.2.(a) and (b) of the Convention are not adequate to achieve the objective stated in of Art. 2,

- Annex I Parties shall strengthen their commitments under Art. 4.2 (a) and (b), and

- all Parties, including non-Annex I Parties, shall continue to advance in the implementation of existing commitments under Art. 4.1.

The distinction is not only the result of a difficult negotiation. It also reflects the fact that most OECD countries already have a satisfactory energy supply, but that is not the situation in developing countries. For developing countries to undertake today the commitments of Art. 4.2 (a) and (b) means to give up the possibility of increasing energy and transport supply to the levels required to development their resources. In most cases that will amount to keeping their people below the minimum levels of education, housing, diet and health.

In the future, further involvement of developing countries will be necessary, but the situation today is that not all developed countries have fulfilled their commitments under the convention: their global emissions are not returning to the levels of 1990 and the financial and technical assistance offered to developing countries is not yet readily available.

This is an issue today in the AGBM negotiations, and some proposals seem to look for the possibility of enlarging the number of Parties that will strengthen their commitments, beyond those included in Annex I. This could lead to an endless debate, because it would depart from the Berlin Mandate. Particular cases of countries which in 1992 were 'developing countries' and are now OECD members perhaps need to be solved among OECD members, but that does not modify the scope of the Mandate.

Delegations to the AGBM have given different emphasis to policies and measures on the one hand, and to quantified emissions limitations and reduction objectives (QELROs) within specified time-frames on the other hand, and this is reflected in statements and proposals.

There is a 'menu approach', under which the protocol or another legal instrument could provide for a detailed listing of policies and measures from which Annex I countries could choose on the basis of their national circumstances; and there is a mandatory approach, under which the new legal instrument would require certain common and/or coordinated policies and measures. One specific proposal is to develop separate lists with mandatory policies and measures, coordinated policies and measures and optional policies and measures.

The concept of policies and measures needs clarification and refining:

- Some delegations have proposed mechanisms for implementing policies and measures, e.g., regulations (Russia); economic instruments, voluntary agreements, education and training (Germany); performance indicators (Australia).
- Some have proposed policy objectives, e.g., reducing emissions from transport, industry, agriculture or forestry (Ministerial Declaration); promoting fuel switching (Japan).
- Others have proposed specific policies, e.g., CO_2/energy tax (Germany); CO_2 emission efficiency targets (Japan); energy efficiency standards (EU).

On the quantified emission limitations and reductions, there is also a mosaic of ideas. It has been proposed that QELROs shall be legally binding (Ministerial Declaration); legally binding but with a certain degree of flexibility for Parties with economies in transition (Germany); and legally binding with substantial safety margins (UK). Most probably all those considerations will be taken into account in the final text.

There are also different views on how greenhouse gases will be covered by targets:

- all of them in a single basket;
- a single target that would initially only relate to a specifically defined list of gases because of the varying degrees of scientific knowledge and data availability with regard to different greenhouse gases;
- different targets for different gases;
- a separate objective to cover reduction of emissions from international bunker fuels.

Flexibility and differentiation are also elements of the negotiations. Perhaps those are the key areas for agreement. We need to be imaginative to facilitate the participation of each and every Party and, at the same time, achieve the objective of the Convention.

Different views have been presented on the time-frames for applications of QELROs. At the third session of AGBM it was noted that some Parties emphasized short- and medium-term goals (2005 and 2010) to promote early action, while recognizing that a longer-term perspective could complement these; and also that other Parties, while recognizing the advantages of short-term targets, were more inclined to set long-term targets to optimize investment decisions.

It has been proposed that Annex I Parties should have individual emission limitations target (Germany), but it was also suggested that a collective emission limitation and reduction objective would be established for Annex I Parties, with differentiated commitments for individual Parties (Australia).

In order to minimize costs it has been proposed that targets cover a number of years, with a kind of 'emissions banking', instead of a flat target for a given year.

Continuing to advance the implementation of existing commitments under Art. 4.1 of the Convention is an integral part of the Berlin Mandate, and this applies to both developed and developing countries.

From the point of view of developing countries, that means producing the first national communications and taking the first approaches to mitigation and adaptation, of course in the context of their own national development programmes. The technical and financial contribution of developed countries to this endeavour is a condition *sine qua non*, for practical reasons, such as lack of experience, know-how and resources in the developing countries, and also for legal reasons in accordance with Art. 4.7 of the Convention.

As you can see, the possibilities are many, and only a wise and balanced combination will allow agreement. Nobody could say that it is easy to reach that agreement. Most probably not all Parties to the Convention will become Parties to the Protocol; some governments will do so only after time, experience and political pressure from public opinion.

Many policies and measures can be adopted at no cost, and even with benefits in some cases, but that will require time, the time necessary to justify new investments and the renewal of capital assets. Adequate timeframes to achieve quantified limitations and reductions can help. But this is not the complete solution. It seems very difficult to avoid all costs.

Sharing the costs could facilitate the adoption of policies and measures, but this should not lead to unfulfilled commitments or the forced accumulation of carbon in certain geographical areas through 'tree farming', or to profits being made from 'quasi-blank cheques' given to those Parties that initially do not have quantified commitments.

Equally, experiences developed in trade with emission permits related to short-life gases should not be applied to long-life gases, such as CO_2, without a careful analysis and assessment of the long-term consequences.

Chapter 9

Emission commitments and flexible implementation mechanisms under a strengthened climate agreement

Berndt Bull

The subject of this year's conference is of immediate importance since the issue of flexible implementation mechanisms, such as joint implementation and emissions trading, will undoubtedly receive increased attention in the deliberations on a strengthened international climate agreement in the year to come. On 8–14 Dec 1996 the Ad Hoc Group on the Berlin Mandate (AGBM) commenced negotiations on the substantial content of a strengthened climate agreement to be adopted at the Third Conference of the Parties (COP-3) to the Climate Convention in Kyoto in December 1997. I would therefore like to consider here the connections between the questions of joint implementation and emission permits trading and the present situation in the climate change negotiations.

The Berlin Mandate prescribes the process for advancing the international response to climate change. An identified priority in this process is the strengthening of the commitments of developed countries (Annex I Parties) after the year 2000, to be laid down in a protocol or another legal instrument. In this context, the Ministerial Declaration at the Second Conference of the Parties in summer 1996 was most encouraging. It demonstrated overwhelming support from most governments for taking serious action against climate change. First of all, the Geneva Declaration gives unequivocal support to the findings of the IPCC as a scientific basis for future actions to limit and reduce emissions of greenhouse gases. Second, consistent with this recognition, the Declaration emphasizes the need to accelerate the negotiations on strengthened commitments and advises '*quantified legally binding* objectives for emission limitations and significant overall reductions within *specified time-frames*' to be set for developed countries (Annex I Parties) in the final agreement. The Geneva Declaration therefore represents an important step towards the adoption of an adequate agreement at Kyoto.

Flexible implementation mechanisms should be a part of a strengthened climate agreement. However, a premise for most of these systems is that initial quantitative emission commitments, or entitlements, have been allocated to the Parties to the agreement.

Structure of emission commitments

The question on how to structure and implement so-called 'Quantified Emissions Limitation and Reduction Objectives' (QELROs) will demand increased attention in the forthcoming negotiations. The rather limited discussions on the allocation of emission commitments so far have indicated two general approaches: a flat-rate approach – implying uniform percentage reductions across all countries – and a differentiated approach. It is my view that a *differentiated approach* is necessary according to the Berlin Mandate. The Mandate, consistent with the Convention, emphasizes that account shall be taken of important differences in national circumstances among the Parties. Clearly, a stringent flat-rate approach will not respond to such substantial differences. It fails on grounds of equity but also on grounds of efficiency – it will reward those countries that are most inefficient and emit most, and will penalize those that have already successfully implemented measures to make their economies more efficient.

In a longer-term perspective, the global challenge we are facing requires commitments and cooperative efforts by *all* Parties. The simplified division of countries applied in the Convention is not sufficient to embrace either the current national differences existing among developed countries, or the differences that will exist among developing countries when they are ready to take on emission commitments too. The importance of reflecting such differences in a strengthened climate agreement has also been underscored by the IPCC Working Group III in its contribution to the Second Assessment Report. It is therefore important that the agreement we draw up today establishes a model that sets us on the right track for the future. An agreement that is structured in an equitable way will enable us to take on and realize more ambitious long-term goals.

It is against this background that Norway has responded to the invitation from the AGBM to submit proposals regarding possible criteria and modalities for differentiation of emission commitments. The Norwegian submission substantiates the reasons why uniform targets across Annex I Parties would be inconsistent with the premises of the Berlin Mandate, and why a differentiated approach can lead to outcomes that are both more equitable and more cost-effective. Furthermore it discusses a set of criteria and indicators to be included in a formula for differentiation. The submission also underlines the need for flexible mechanisms to support the implementation of the emission commitments.

In order to facilitate completion of the work of the AGBM by the third session of the Conference of the Parties, a thorough discussion of the elements and preconditions set by the Mandate remains indispensable. Making the premises for emission commitments operational in a protocol or another legal instrument has been and will continue to be a key issue for the AGBM. I hope that the Norwegian submission will be a useful contribution to these important discussions in the forthcoming sessions.

Flexible implementation mechanisms

While equity and cost-efficiency are critical concerns to be taken into account when setting and allocating targets, cost-efficiency must also be duly considered when formulating policies and measures to achieve these targets. Thus an important aim is that, as far as possible, a protocol or another legal instrument should be cost-efficient across greenhouse gases, sectors and countries to achieve the environmental objectives at the least overall cost to the Parties. Any system which distributes emission commitments among Parties has to allow flexible implementation to ensure cost-effectiveness. It is my opinion that, independent of how the emission commitments are initially distributed among Parties, both joint implementation and emission trading can, if designed in a proper way, contribute to significant cost reductions and thus to more ambitious environmental goals.

Joint implementation/activities implemented jointly
Norway has for a long time advocated the concept of *joint implementation*. We therefore welcomed the launching of a pilot phase for activities imple-

mented jointly (AIJ) in Berlin in 1995. The government has renewed its support for the development of reporting criteria and crediting options on AIJ. Norway has so far co-funded two AIJ pilot projects – in Poland and in Mexico – in cooperation with the World Bank, UNEP and UNDP. The project in Poland involves the conversion of several coal-fired power plants to gas, while that in Mexico promotes high-efficiency lighting in two cities. New projects in Costa Rica and Burkina Faso have recently been decided on, and we are in the process of identifying possible AIJ projects in eastern Europe and the Baltic countries as well. In addition to bilateral activities, we have signed a three-year agreement with the World Bank on cooperation on common AIJ activities to catalyse opportunities and maximize learning. This type of cooperation may be helpful in *reducing transaction costs*, such as those related to identification, implementation, monitoring and verification. Finally, the Nordic countries are involved in a joint effort to investigate the climate change effects of ongoing environmental projects in eastern Europe and the Baltic countries together with the Nordic Environmental Financing Cooperation (NEFCO).

Norway will continue to participate in pilot projects on AIJ. One principal aim in our present engagement in AIJ projects is to demonstrate workable solutions, show the risk components of such activities and provide examples of institutional frameworks. In general, projects on fuel switching and energy efficiency may seem preferable to projects on enhancement of sinks, especially because of measurement problems related to the latter. Moreover, it is essential to develop the potential role of the private sector and mobilize its funds. The private sector must, in this respect, be willing to consider the pilot period as an opportunity for investments in knowledge and practical experience without achieving credits for emission reductions.

I feel confident that, when more generally applied, and in a longer-term perspective, joint implementation will be a necessary instrument in achieving the ultimate objective of the Convention. However, a significant challenge will be to find ways of reducing the transaction costs related to identification, implementation, monitoring and verification. In our view, the agreement we are now negotiating should allow for joint implementation between Annex I Parties. The possibilities for such arrangements in the upcoming agreement should therefore be further investigated. The door should also be kept open for similar arrangements with non-Annex I Parties, provided such cooperation is in accordance with national aims and priorities in receiving countries and implemented in a transparent way, according to agreed criteria. This conference should contribute valuable insight into how such systems might be designed and implemented.

Emission trading

The second concept focused on at this conference, *tradeable emission permits* or quotas, is also highly interesting, although international experiences with such systems are still few. Emissions trading systems certainly have the potential to enhance the overall cost-effectiveness of international environmental agreements, provided that monitoring and enforcement are adequate. In this respect, the experience gained from the US tradeable sulphur permit scheme might provide valuable lessons and insight for the design and implementation of corresponding systems at international level within the field of climate change as well. Still, with the variations in domestic regulatory policies existing among Parties, we are in need of additional practical experiences from more countries. I would like to mention that Norway is currently exploring new instruments to ensure that our SO_2 emissions remain below the ceiling set in the Second Sulphur Protocol. At present, our most important instruments are a tax on the sulphur content of mineral oils and a licensing system for industry emissions. There are basically two alternatives for our future approach:

(1) Extending the sulphur tax, primarily to cover coal and coke in industrial processes, so that the content of sulphur in these products is taxed in line with sulphur in mineral oils.
(2) Establishing a system of tradeable sulphur quotas which will encompass all major sources of SO_2 emissions.

We are developing the specific elements in a tradeable permit system in close contact with the relevant industries. Which alternative we choose will probably be decided in autumn 1997, but I hope that experiences gained from any trading scheme might be carried forward to systems for control of GHG emissions.

We also need further intellectual efforts on legal and institutional aspects and fundamental design parameters for international trading schemes. I find both the UNCTAD project on pilot emission trading programmes and the project conducted by the Annex I Experts Group on the FCCC (supported by the OECD and the International Energy Agency (IEA) valuable and important as a basis for further exploration and possible input to the present negotiations. It is vital that such trading systems are kept as simple as possible, that they are practicable, transparent and feasible. The system should also be able to function without the establishment of a large international bureaucracy to monitor and enforce the trading.

As with joint implementation, we need practical experience with systems of emission trading at the international level. As mentioned before, it is in my view necessary for the agreement reached in Kyoto to include flexible market mechanisms such as emissions trading. While the content and distribution of emission commitments will probably attract much of the attention in the climate change negotiations in the near future, further effort is also needed on development of flexible implementation mechanisms to enhance the cost-effective implementation of the forthcoming commitments.

Chapter 10

Opportunities and concerns in the developing world: the case of China

Li Yun

Carbon and sulphur emissions in China

At present, China has the third biggest energy system in the world, and its total primary energy output is second only to that of the United States and Russia. Raw coal output ranks first in the world, crude oil output fifth, and electricity generation second. China is one of the few countries whose energy structure is dominated by coal. In 1994, the total primary commercial energy consumption was 1227.37 million tonnes of coal equivalent (Mtce),[1] among which coal accounted for 75.0 per cent, oil 17.4 per cent, natural gas 1.9 per cent, hydropower 5.3 per cent, and nuclear power 0.4 per cent.

Because of the huge energy consumption and coal's domination of the energy structure, large emissions of sulphur dioxide and carbon dioxide are inevitable. In 1994, China's SO_2 emissions (not including town and village industry) totalled 18.25 Mt, of which coal combustion accounted for about 90 per cent. The major sources of coal combustion were electricity generation (accounting for about one-third of the total), industrial boilers (about one-fifth), and domestic consumption (about one-tenth).

According to research by the Energy Research Institute of State Planning Commission (ERI of SPC), in 1990 CO_2 emissions from energy and industrial sources totalled about 600 MtC.

The research by ERI also indicates that if implementation of China's current energy policies continues, coal's proportion in primary energy consumption will be 71.6 per cent, 68.3 per cent and 63.1 per cent in 2000, 2010 and 2020 respectively. If annual energy conservation rates are 4.3 per cent, 4.0 per cent and 3.2 per cent respectively, the annual growth rate of

---

[1]   764.78 Mtoe (United Nations, *Energy Statistics Yearbook*, 1996). The conversion factor from Mtce to Mtoe is about 1.45.

SO_2 emissions will be 3.9 per cent, 2.7 per cent, and 1.3 per cent, and the annual growth rate of CO_2 emissions will be 4.7 per cent, 2.9 per cent and 1.8 per cent respectively. To draw analogous conclusions, if no effective mitigation measures are adopted, by the year 2020, SO_2 and CO_2 emissions from energy activities will amount to 35.5 Mt and 1640 MtC, meaning that China will rank respectively first and second in the world.

China's record in controlling carbon and sulphur

The large quantities of SO_2 emissions have resulted in severe atmospheric pollution and large areas of acid rain. The Chinese government takes SO_2 emissions seriously and is positively adopting measures to control and mitigate them. Global climate change caused by CO_2 and other greenhouse gases will seriously affect the environment, as well as social and economic development, and the Chinese government is therefore equally concerned to control and mitigate carbon emissions.

At the same time as developing its economy, China adopted many direct and indirect measures for mitigating carbon and sulphur emissions, and has achieved good results. The principal measures include increasing energy efficiency, developing clean coal technologies, expanding the utilization of new and renewable energy, afforesting and controlling population growth.

During the past five years, the Chinese government has pursued the principle of laying equal stress on energy development and energy saving, given efficient utilization of energy and raw materials a high priority in its energy policy, and devoted major efforts to developing energy-saving measures. In all this it has achieved remarkable success. In the Eighth Five-Year Plan period, the value of total domestic output increased by an average of 12 per cent annually, while that of primary commercial energy output increased by an average of 3.6 per cent annually. The elasticity of energy consumption was 0.47, the annual average energy conservation rate reached 5.9 per cent in accordance with the calculation of GDP, and total energy savings were 358 Mtce. This was equivalent to a decrease of about 50 Mt in CO_2 emissions and about 2 Mt in SO_2 emissions.

In 1995, China's total capacity for coal washing was 380 Mt; coal for washing totalled 280 Mt, or 22 per cent of coal output. This removed 1.01 Mt sulphur, equivalent to 2.02 Mt SO_2. To date, utilities of about 1.5 GW capacity and large and medium-sized coal-fired industrial boilers are equipped with desulphurization facilities, most of which are imported from abroad and are in the demonstration phase. The relevant domestic departments are arranging for experts to learn about various desulphurization technologies. As regards other coal-fired facilities, no appropriate mitigation options for sulphur dioxide emissions are widely available for commercial use.

Since the 1980s, China has been actively developing and utilizing new and renewable energy sources, and has achieved some improvements in energy structure and mitigation of CO_2 and SO_2 emissions. In 1994, new and renewable energy supplies totalled 318.7 Mtce, of which biomass accounted for 248.0 Mtce, hydro power 68.7 Mtce, and new energy 2.0 Mtce. This new and renewable energy accounted for 21.6 per cent of primary energy consumption (commercial and non-commercial combined).

To enhance the environment, the Chinese government has undertaken large-scale afforestation projects, which have led to considerable success in carbon and sulphur control. This includes the world-renowned 'Three North Shelter Forest Programme', which is described as 'the first ecological engineering in the world', in which the total afforested area was 18.5 Mha; total afforestation in the 'Middle and Upper Reaches of the Yangtze River Shelter Forest Programme' reached 5.5 Mha. In addition, the 'Mount Taihang Afforestation Programme', 'Coastal Shelter Forest Programme' and other large-scale afforestation programmes are proceeding smoothly. During the Eighth Five-Year Plan period, the cultivation of forest resources was further speeded up. The total completed afforestation was 29.5 Mha, with voluntary planting of 12.1 billion trees; and twelve provinces have eliminated the development of afforestated land. At present, 263 Mha of land in China is afforested; natural forest area totals 134 Mha, and man-made forest 34 Mha. The forest-cover rate is 13.9 per cent, growing stock 11.8 billion cubic metres, forest growing stock 10.1 billion cubic metres, forest biomass 20.6 billion tons, and carbon content in forests 10.3 billion tons. At the end of the 1980s and the beginning of 1990s, both forest area and growing stock increased in China. Forest land became a weak sink of CO_2, and the current amount of net fixed carbon is about 50–86 MtC.

The Chinese government is trying hard to raise its people's awareness of afforestation and to increase the relevant investment. It is attempting to expand the afforested area by 4–5 Mha per year, especially fast-growing and rich harvest forest. By the year 2010, China will have afforested all suitable barren mountains, and the forest-cover rate will reach 17 per cent. Through this large-scale afforestation programme, it is estimated that 80–150 Mt carbon will be absorbed per year, reducing total CO_2 emissions in China by 10–15 per cent.

To reduce energy demand and environmental pollution caused by population growth, the Chinese government has successfully been pursuing a national policy of birth control and population control since 1973. At the beginning of the 1970s, the natural growth rate of the population was as high as 2.3 per cent. This had fallen to 1.439 per cent in 1990, and was further reduced to 1.055 per cent by 1995. This was much lower than the natural growth rate of other developing countries, and also than the world average. On the basis of the natural growth rate of 1970, by the year 1995, the population would have expanded by 360 million. During this period, average per capita energy consumption increased from 0.35 tce in 1970 to 1.05 tce in 1995. If the average per capita energy consumption was 0.7 tce, the corresponding energy conservation was as much as 250 Mtce. On the basis of the unit energy carbon emission intensity in 1990 (which was 0.60 tC/tce) and unit energy SO_2 emission intensity (which was 1.62 tSO$_2$/100tce), over the 25 years of the birth control policy the emissions of carbon and SO_2 were reduced, respectively, by 150 Mt and 4 Mt.

China will continue to control its population increase, and to implement the basic national policy of birth control. It will try to achieve its population control targets of under 1.3 billion in 2000, under 1.4 billion in 2010, and about 1.5 billion in 2050. If one calculates the population according to the natural growth rate of 1995, by the year 2050 the corresponding population reduction will be as much as 650 million. During the next 50 years, if the average per capita energy consumption is 2 tce, unit energy carbon emission intensity 0.5 tC/tce and unit energy SO_2 emission intensity 1.5 tSO$_2$/100tce, then the accumulated reduction emissions of carbon and SO_2 will amount to 650 Mt and 19.5 Mt respectively – equivalent to the current annual carbon and SO_2 emissions.

Although China is the world's third largest energy consumer because of its large population, per capita energy consumption is very low. In 1994, per capita commercial energy consumption was only 1.024 tce, which was only 50 per cent of the world average. Average per capita domestic electricity consumption was only 73 kWh – only 2.2 per cent of the US figure – and there are still 100 million people without access to electricity. Given that per capita GNP is less than US$500, the campaign to reduce SO_2 emissions from energy activities will not only ameliorate the quality of China's environment, but also have a positive impact on the global environment. As China and other developing countries progress socially and economically, their energy consumption and hence carbon emissions will obviously increase. The Chinese government firmly supports international action to control and reduce carbon emissions, and, as always, will continue to contribute towards this.

Concerns in China and other developing countries

Consensus has gradually been reached on the adverse effects on humans of environmental pollution and global climate change. Controlling these effects is one of the important components of sustainable development strategy. On the one hand, although the current per capita SO_2 and carbon emissions in developing countries are still very low, energy consumption levels and hence emissions of SO_2 and other pollutants, CO_2 and other greenhouse gases will increase. On the other hand, most of the developing countries are more vulnerable than developed ones to environmental pollution and climate change, and it is therefore more necessary and more urgent for developing countries to control them.

However, developing countries face many difficulties in their attempts to reduce carbon and sulphur emissions, particularly because their reduction technologies and funds are limited. China and other developing countries hope that developed countries will urgently transfer technologies and provide capital, especially in areas such as energy conservation, clean energy, and new and renewable energy.

Because most past and current greenhouse gas emissions and environmental pollution stem from developed countries, it is their duty to provide technologies and capital to developing countries. As the UNFCCC points out:

> The developed country Parties and other developed Parties included in Annex II shall take all practicable steps to promote, facilitate and finance, as appropriate, the transfer of or access to, environmentally sound technologies and know-how to other Parties, particularly developing country Parties, to enable them to implement the provisions of the Convention. In this process, the developed country Parties shall support the development and enhancement of endogenous capacities and technologies of developing country Parties.

> … The extent to which developing country Parties will effectively implement their commitments under the Convention will depend on the effective implementation by developed country Parties of their commitments under the Convention related to financial resources and transfer of technology and will take fully into account that economic and social development and poverty eradication are the first and overriding priorities of the developing country Parties.

To mitigate carbon and sulphur emissions, China and other developing countries will undertake large-scale control measures, and there will be an important market in this field. Therefore, it will benefit both developing and developed countries if the latter provide technologies and funds for the former. It is a common duty for humans to improve the environment, and the improvement of the global environment will benefit all.

Investment opportunities in China

In 1994, the Chinese government approved China's 'Agenda 21',[2] a white paper on population, environment, and development in the twenty-first century. To realize the objectives formulated in Agenda 21, a Priority Programme has been drawn up, with the projects divided into categories based upon the programme areas. These projects will serve as the

---

[2]   *China's Agenda 21*, China's Environmental Science Publishing House, 1994.

fundamental means for implementing Agenda 21. At the same time, to implement the commitment to the UNCED in 1992, the Chinese government is taking practical steps to fulfil the strategy for sustainable development. The Priority Programme will be embodied into the medium- and long-term plans for national economic and social development, particularly in the formulation of the Ninth Five-Year Plan (1996–2000) and the plans for 2010 at various levels, based on domestic resources and international support.

Organization and implementation of the Priority Programme

- The projects selected in the Priority Programme will be incorporated into the medium- and long-term national economic and social development plans for implementation, after they have been appraised and approved in accordance with certain procedures.
- With UNDP assistance, the State Planning Commission, the State Science and Technology Commission, the State Economic and Trade Commission and the National Environmental Protection Agency will guide and coordinate the implementation of the priority projects. The Administrative Centre for China's Agenda 21 (ACCA21) will undertake the routine administration.
- The Priority Programme will be carried out through many forms of international cooperation, such as multilateral or bilateral grants, loans, foreign investment, joint ventures, build-operate-transfer (BOT) agreements, etc.
- The priority projects will be executed jointly by national implementing agencies and associated international partners, and with wide participation from all parts of society..
- The Priority Programme will be executed in a rolling and flexible way. The priority projects that have been and are being implemented will be evaluated and adjusted in due course. The remaining proposals which have not been covered in the first-tranche Priority Programme, and newly submitted project proposals, will be put into the priority project inventory. Details will be available from the ACCA21 and they will gradually be incorporated into the national economic and social development plans, if possible. Continuous efforts will be made to seek international cooperation and assistance to promote the implementation of these projects.

Priorities concerning carbon and sulphur control

- *Priority 3 – clean production and environmental protection industry.* This includes the management of clean production, the introduction and demonstration of clean technologies in principal industrial sectors and enterprises, desulphurization, dust removal, the development of an environmental protection industry and the construction of an industrial park for environmental science and technology, etc.

- *Priority 4 – clean energy and transportation.* This includes clean coal technologies, increased energy efficiency, utilization of renewable energy sources, modern transport planning, etc.

- *Priority 5 – conservation and sustainable utilization of natural resources.* This includes prevention and control of soil erosion, conservation of wetland resources, natural resources accounting, the establishment of a monitoring network for ecology and the environment, reclamation of wastes and mine tailings, etc.

- *Priority 6 – environmental pollution control.* This includes waste water treatment and recycling, safe management of hazardous wastes and toxic chemicals, the treatment and disposal of radioactive wastes, acid rain control, etc.

- *Priority 9 – global change and biodiversity conservation.* This includes global concerns such as climate change, conservation of biodiversity, prevention and control of desertification, etc.

Financial inputs

The implementation of China's Agenda 21 will depend mainly upon China's own resources, but at the same time China will actively seek international cooperation and assistance in this field. Roughly speaking, the Chinese inputs will cover about 60 per cent of total funding and the remainder is expected to come from the international community. In line with input means, the priority projects will be basically divided into three types: assistance, cooperation, and investment. China is willing to discuss methods of inputs and the amount of funding in detail with international partners who are interested in the Priority Programme for China's Agenda 21.

In March 1996, China approved an Outline of the Ninth Five-Year Plan and long-term objectives to the year 2010 for national economic and social development. The objective for environmental protection is as follows:

> By the end of this century, to try to basically control the worsening trend towards environmental pollution and ecological damage; to improve the environment of some cities and regions to a certain extent; by the year 2010, to basically change the situation of ecological and environmental deterioration, to improve the environment of urban and rural areas by a noticeable extent.

The objective for the energy industry is:

> The energy industry should be suitable for the requirements of national economic growth, and the bottleneck should gradually be eased. China's total primary energy output will increase from 1.24 billion tce in 1995 to 1.35 billion tce in 2000. Both development and conservation will be insisted on, with initial priority given to conservation. Major effort will be devoted to adjusting the energy production and consumption structure. Advanced technologies will be popularized, and energy production efficiency increased. Energy development and treatment of the environment will be synchronized, and energy products continuously rationalized.

To achieve the objectives mentioned in the Outline, different sectors have formulated relevant objectives and measures. The following are some examples.

(a) To save energy and increase energy utilization efficiency in the Ninth Five-Year Plan period, the State Economy and Trade Commission will focus on giving an impetus to the following six demonstration projects:

- A clean coal combustion project, including the perfection of a circulating fluid-bed boiler, increasing integral operating efficiency, industrial briquets, and joint thermal, electricity and gas production;
- A power-saving project of electricity and electronics, including a high-efficiency fan and water pump and speed control technology;

- A green lighting project, including high-efficiency electric light source, appliance and light control system;
- A project for recovering excess energy (such as surplus pressure and heat) from industrial processes such as dry coke processing, blast furnaces, and cement kilns, for example electricity co-generation using waste heat;
- A raw material conservation project, including new technology for lower usage and non-use of cutting and surface treatment, and a new technique of continuous extrusion.

(b) The Ministry of Electric Power Industry formulated the objectives for thermal power plants: in the Ninth Five-Year Plan period, five desulphurization demonstration projects for thermal power plants will be completed and the relevant experience will be collated. Some of the thermal power plants located in the Acid Rain Control Zone and SO_2 Pollution Control Zone will have desulphurization equipment installed. With the support of economic policies, the installed capacity of desulphurization equipment will amount to 10 GW by the year 2000.

(c) The State Science and Technology Commission identified eight priorities for development:

- Solar energy utilization technology;
- Solar photovoltaic technology;
- Large-scale wind power generation and relevant technology;
- Biomass technology;
- A survey and evaluation of high-temperature geothermal energy resources and development of power generation and heat supply technologies;
- Ocean energy technology;
- Hydrogen energy technology;
- Fuel cell technology.

(d) With the enactment of the new revised Law on Air Pollution Control and Prevention, the Ministry of the Coal Industry is actively taking the following counter-measures during the Ninth Five-Year Plan period to reduce the sulphur content in coal through the incorporation of environmental considerations in coal production planning:

- *Adjustment of the layout of coal production and construction.* The target is to increase the proportion of coal output in regions with low sulphur content to total nationwide output from 70 per cent in 1995 to 73.3 per cent in 2000 through the accelerated construction and development of mines producing steam coal with a low sulphur and ash content (in western and central China), together with reduced investment in state-owned mines in several southern provinces that have coal with a high sulphur content, and restricted production from mines producing coal with a high sulphur and ash content.
- *Expansion of coal washing capacity for newly constructed mines.* The target is to increase the proportion of coal for washing to total coal output nationwide from 22 per cent in 1995 to over 30 per cent in 2000, with the most dramatic increase in the region producing coal with a high sulphur content.
- *Installation of coal washing plants which process coal with a high sulphur content.* About 19 coal washing plants will be put into operation during the Ninth Five-Year Plan.
- *Modernization of coal washing plants which process coal with a high sulphur content.* About 16 coal washing plants will be retrofitted during the Ninth Five-Year Plan.

(e) The State Development Bank (SDB) said that it would continue to fully support the development of both primary and secondary energy industry in China by optimizing its credit management and service, broadening its cooperation with international financial circles and sustaining loans and financial services to joint-venture or energy cooperation projects. In view of this target, SDB will actively increase the use of foreign capital, export credits, foreign commercial borrowings, on-lending and bond issuance in international markets.

To absorb more foreign capital, and use foreign capital more rationally and more effectively, China has formulated many relevant laws and policies.

It sincerely welcomes proposals from other countries to promote further cooperation on the treatment of the environment, climate change mitigation, carbon and sulphur control, and other relevant projects.

References

State Economic and Trade Commission, China Energy Research Society, *China Energy Annual Review,* 1996.

Shi Dinghuan, *Research on the New and Renewable Energy and its Development in China,* State Science and Technology Commission, 1996.

United Nations Framework Convention on Climate Change, UNEP/WMO, 1994.

Wu Baozhong, *Strategies to Control SO$_2$ Pollution Related to Energy in China,* National Environmental Protection Agency, 1996.

Yang Ziwei, *Energy Saving Measures and International Cooperation in China's Ninth Five-Year Plan Period,* State Economic and Trade Commission, 1996.

Chapter 11

The US Initiative on Joint Implementation

Robert K. Dixon[*]

Abstract

More than 150 countries are now Party to the United Nations Framework Convention on Climate Change, which seeks, as its ultimate objective, to stabilize atmospheric concentrations of greenhouse gases at a level that would prevent dangerous human interference with the climate system. As a step toward this goal, all Parties are to take measures to mitigate climate change and to promote and cooperate in the development and diffusion of technologies and practices that control or reduce emissions and enhance sinks of greenhouse gases.

In the US, efforts between countries or entities within them to reduce net greenhouse gas emissions undertaken cooperatively (Joint Implementation) hold significant potential both for combating the threat of global warming and for promoting sustainable development. To develop and operationalize the JI concept, the United States launched its Initiative on Joint Implementation (USIJI) in October 1993, and designed the programme to attract private-sector resources and encourage the diffusion of innovative technologies to mitigate climate change.

The USIJI provides a mechanism for investments by US entities in projects to reduce greenhouse gas emissions worldwide and has developed a set of criteria for evaluating proposed projects for their potential to reduce net GHG emissions. The criteria are designed to identify and allow the USIJI to 'accept' projects that:

[*] This report summarizes the accomplishments of the USIJI Secretariat and the manuscript is offered as an invited USIJI contribution to the conference. A number of individuals contributed to the report, including Dr E. Holt, Ms J. Leggett, Dr J. Pershing and members of the USIJI Secretariat. This manuscript has not been subject to technical or policy review and does not reflect the view(s) of the US government nor of any intergovernmental body.

- Support the development goals of the host country while providing greenhouse gas and other environmental benefits.
- Produce measurable reductions in addition to any likely without the project.
- Can be monitored and tracked.
- Will not result in net greenhouse gas emissions elsewhere or otherwise have significant secondary environmental impacts.
- Have enduring impact.

To date, the USIJI has received more than 50 project proposals. Of these, 22 have been accepted. These projects represent a diverse set of innovative technologies and practices in six countries, and include projects developing renewable energy sources such as solar, biomass, and hydro-electric power, and land-use change projects leading to better forest management, reforestation and afforestation.

Aggregating preliminary estimates presented to the USIJI by project developers suggests that cumulative net emission reductions as a result of these projects are expected to be nearly 30 million tonnes of carbon (MtC) equivalent. Although the USIJI does not certify project estimates prospectively, it does set forth provisions for monitoring and verifying emissions reductions as they occur. Furthermore, accepted projects, when fully implemented, are expected to lead to significant financial and technical investments in host countries.

Additional proposals considered by the USIJI include submissions for projects in 12 countries and in such other technical areas as methane reduction from livestock and waste treatment. Eight of these proposals were withdrawn and another ten were not accepted. However, 18 of these proposals have been placed 'in development' and will receive limited technical support in order to assist them in fully meeting USIJI criteria for acceptance.

To test the USIJI criteria and to provide input into the international pilot phase, the United States intends to promote the development of other 'acceptable' projects, and to seek additional information on the experience of individual developers during project implementation. As a supplemental effort, the USIJI seeks to assist countries in developing their national joint implementation programmes and to this end has not only developed a domestic outreach effort, but also sponsors regular international work-shops, produces a USIJI newsletter, and maintains a home page on the World Wide Web.

Background

At the 1992 Earth Summit in Rio de Janeiro, the United States joined more than 150 countries in signing the United Nations Framework Convention on Climate Change. Parties to the Convention recognized that human activities have contributed substantially to increases in atmospheric con-centrations of greenhouse gases, and that this trend poses a serious threat to the earth's climate system. The ultimate objective of the FCCC, as called for in Article 2, is the 'stabilization of greenhouse gas concentration in the atmosphere at a level that would prevent dangerous anthropogenic interference with the climate system'.

The concept of joint implementation was introduced early in the nego-tiations and was formally adopted in the FCCC text. Article 4(2)(a) of the Convention explicitly provides for Parties to meet their obligation to reduce greenhouse gas emissions 'jointly with other Parties', that is, through joint implementation activities. JI has been used to describe a wide range of possible arrangements between interests in two or more countries, leading to the implementation of cooperative development projects that seek to reduce, avoid, or sequester greenhouse gas emissions.

At the First Meeting of the Conference of the Parties (COP-1) in Berlin in March and April 1995, the Parties addressed 'decisions regarding crite-ria for joint implementation' (Article 4). At this meeting, the Parties deter-mined that there would be an initial pilot phase of JI referred to as 'Activities Implemented Jointly' (AIJ) and that during this phase of JI, 'no credits shall accrue to any Party as a result of greenhouse gas emissions reduced or sequestered during the pilot phase …'. The pilot phase ends no later than the year 2000.

COP-1 further decided that the Subsidiary Body for Scientific and Technological Advice (SBSTA), coordinating with the Subsidiary Body on Implementation (SBI), would 'establish a framework for reporting … the global benefits and the national economic, social, and environmental impacts as well as any practical experience gained or technical difficulties encountered in AIJ under the pilot phase'. The SBSTA adopted a reporting framework at its second meeting in March and April 1996. SBSTA invited the Parties to report on AIJ through the FCCC Secretariat. SBSTA and SBI will produce a synthesis report from these submissions which will be considered by the Conference of Parties on an annual basis. These reports will also form the basis for 'improving the reporting framework and for addressing methodological issues'.

Why Joint Implementation?

Greenhouse gas emissions are rising rapidly. The most cost-effective options for addressing this problem often exist in developing countries, or countries with economies in transition, as they restructure or expand their infrastructure. Because costs of reducing or sequestering emissions of greenhouse gases vary among countries, and all such emissions have the same effect on global climate regardless of where they are emitted, joint implementation offers the opportunity to reduce emissions at a lower global cost than would be possible if each country acted alone. The Parties to the Convention recognized that '... policies and measures to deal with climate change should be cost-effective so as to ensure global benefits at the lowest possible cost ...'. As such, the FCCC stated that '[e]fforts to address climate change may be carried out cooperatively by interested Parties'.

Concerns have been raised by some countries that joint implementation is a means for industrialized countries to transfer their environmental problems to developing countries. However, the purpose of JI is not to provide a mechanism for industrialized countries to export emissions of greenhouse gases. Instead, the COP recognized that JI holds significant potential for lowering the cost of combating the threat of global warming, while also contributing to sustainable development. Furthermore, projects developed under JI can result in technology choices which meet the development objectives of host countries while also achieving the environmental objectives of the FCCC.

In addition, JI can influence technology choices in developing countries, and countries with economies in transition, as they build infrastructure.

Net private and public capital flows to developing countries are approaching $250 billion per year. The goal of joint implementation is to affect these already significant capital flows by increasing the number of environmentally friendly projects supported by these funds.

The US goal through the pilot phase of JI is to gain experience and knowledge which can be used as a basis for the post-pilot phase programmes. This may best be accomplished by:

(1) encouraging rapid development and implementation of cooperative, mutually voluntary, cost-effective projects aimed at reducing or sequestering emissions of greenhouse gases, particularly projects promoting technology cooperation with and sustainable development in developing countries and countries with economies in transition to market economies.
(2) promoting a broad range of projects to test and evaluate methodologies for measuring, tracking, and verifying costs and benefits.
(3) establishing an empirical basis to contribute to the formulation of international criteria for joint implementation.
(4) encouraging private-sector investment and innovation in the development and dissemination of technologies for reducing or sequestering emissions of greenhouse gases.

Joint implementation activities provide benefits for partner country participants and for the global community as a whole. The benefits at the global level include reducing the overall global cost of greenhouse gas emissions reductions while promoting sustainable development.

Benefits accruing to participants within host countries include:

• *Technology transfer.* Encourages private-sector diffusion of innovative technologies that can help meet host country development priorities while reducing or sequestering greenhouse gas emissions.
• *Investments.* Expands investments in technologies and projects that reduce greenhouse gas emissions while contributing to overall host country development objectives.
• *Local environmental and human health benefits.* Produces other local

environmental and human health benefits by preventing or reducing air, water or soil pollution, and/or by contributing to more sustainable use of natural resources.

- *Local economic benefits.* Generates local economic benefits which may include training, construction of new or improved facilities, public participation in projects, provision of new energy services.
- *Promotion of sustainable development.* Encourages additional private-sector investment in the development and dissemination of technologies and practices that contribute to sustainable development while reducing or sequestering greenhouse gas emissions.
- *Influence on the future of JI.* Provides participants with an opportunity to influence the direction and structure of JI beyond the pilot phase by demonstrating the potential for international collaboration to resolve environmental problems.

Benefits to participants outside the host countries include:

- *Market access.* Provides entrée into energy and environmental markets in host countries. Participants may also be eligible for host country assistance in terms of relaxed permitting, reduced import restrictions, local content requirements, and/or tariffs.
- *Lowering the cost of 'green' technologies.* Enhances the competitiveness of 'green' technologies by accelerating application worldwide and further reducing the cost of production.
- *Enhanced prospects for financing.* Expands partnership opportunities by providing greater visibility and credibility to the potential project which can, in turn, increase the depth of creditworthiness associated with the project.
- *Reduction of risk.* Offers greater security of investment in foreign countries.
- *Expansion of knowledge of the JI option.* Provides participants with an opportunity to influence the direction and structure of JI beyond the pilot phase by demonstrating the potential for international collaboration to resolve environmental problems.
- *Recognition.* Demonstrates participants' commitment to reduce the

threat of climate change and contribute to sustainable development.

- *Record of reductions.* Establishes a public record of emissions-reducing activities.
- *International credibility.* Establishes a track record in international markets by working with governments, businesses, and organizations in foreign countries.

The US Initiative on Joint Implementation

In October 1993, President Clinton announced the US Climate Change Action Plan, which set forth a series of measures designed to return US greenhouse gas emissions to 1990 levels by the year 2000. This plan relied on domestic actions alone. However, recognizing the enormous potential for cost-effective greenhouse gas emission reductions in other countries, the administration also called for a pilot programme to help establish an empirical basis for considering cooperative approaches such as joint implementation, and thus help realize the potential of domestic and international strategies to both combat the threat of global warming and promote sustainable development.

On 17 December 1993, draft ground rules for the USIJI programme were published in the Federal Register for public comment. The final ground rules, including a discussion of the specific comments received, were published by the Department of State in a Federal Register notice on 1 June 1994. They describe the purpose of the pilot programme, outline the timeline for evaluation and reassessment of the programme, define eligibility criteria for domestic and foreign participants, establish an Evaluation Panel to assess projects submitted for inclusion in the USIJI, and delineate the criteria for acceptance of a project submission into the USIJI portfolio.

USIJI is among the first and most developed joint implementation pilot programmes worldwide. Its international outreach activities and workshops (attended by several hundred potential participants from approximately fifty countries) have contributed to international understanding of joint implementation and its broad acceptance by Parties to the Framework Convention on Climate Change.

The USIJI is overseen by an Interagency Working Group (IWG) that has the primary responsibility for policy development and criteria used for project acceptance. In turn, ultimate responsibility for project approval and the process for proposal identification, development and evaluation rest with the Evaluation Panel. The panel has one member each from:

- Department of Energy
- Environmental Protection Agency
- Agency for International Development
- Department of Agriculture
- Department of Commerce
- Department of the Interior
- Department of State
- Department of the Treasury.

The Panel is co-chaired by the members from the Department of Energy and the Environmental Protection Agency. A Secretariat supports day-to-day operation of the USIJI programme. Technical experts are drawn from a wide variety of organizations to assist the Secretariat in the proposal review process and to provide technical assistance.

As indicated in the previous section, anthropogenic emissions of greenhouse gases are rising rapidly. The cost of mitigating these emissions will be substantial, and the potential magnitude of the problem is such that government funds and standard technologies alone are likely to be inadequate. Greater resources and innovative technologies need to be brought to bear on the problem. To this end, the private sector needs to be engaged as a major participant. A key goal of the USIJI programme is to influence the technological choices associated with the already substantial private capital flows to developing countries.

Acceptance into the USIJI programme provides US firms with both visibility and credibility. It is hoped that USIJI will result in an increasing number of projects that complement the development goals of the host country and promote the sustainable use of natural resources.

The USIJI is designed to meet general concerns of the Parties, including whether projects:

(1) produce measurable reductions;
(2) are funded independent of or with resources in addition to the FCCC financial instrument or Official Development Assistance (ODA);
(3) measure and track net emission reductions achieved;
(4) ensure that reductions in one place do not give rise to increases in another;
(5) ensure that reductions will not be lost or reversed through time.

Central to the programme is the establishment of criteria designed to meet these concerns and a project evaluation process where the criteria are applied. These aspects of the USIJI programme are considered in subsequent sections.

Project criteria

Projects accepted into the USIJI programme are evaluated against nine criteria and four other areas of consideration. These criteria are intended to identify projects that support the development goals of the host country while providing greenhouse gas benefits beyond those that would occur in the absence of the joint implementation activity. The criteria have been formulated to ensure that projects accepted into the programme will produce real, measurable net emissions reductions.

The Evaluation Panel is responsible for approving or rejecting project submissions for inclusion in the USIJI programme on the basis of specific criteria. The Panel considers how a project measures against all criteria, as well as how it contributes to the purposes of the pilot programme. While failure on any single criterion could keep a project from being approved, the Panel may find relatively poor performance on one criterion to be outweighed by excellent performance on another. Similarly, if a project's performance on all criteria is seen as only barely acceptable, it may not be approved by the Panel.

The application of criteria is also balanced by the goal of the USIJI project to promote a broad range of projects to test and evaluate methods to measure, track and verify costs and benefits of accepted projects. In addition, the criteria are also being tested for appropriateness in selecting projects that will produce real, measurable results. As such, there has not been a single, rigid approach to the application of these criteria, but instead the Evaluation Panel has remained flexible in the interpretation and application to each project. The development of criteria is seen as an evolving process, particularly during this pilot phase. The criteria and other considerations used by USIJI to screen proposals are discussed below.

Acceptable to host country

Proposals must provide written notification from the designated ministry or other entity of the host country national government that the project is acceptable for inclusion in the USIJI programme. Such certification is necessary to ensure that the host country is familiar with the project and that proposed activities are considered to be consistent with the national development objectives. In some countries, a single ministry has been designated to perform this function. Such is the case in Costa Rica, where the Minister of Natural Resources and Energy (MINAE) is the host government-approved signatory for proposed projects being considered by USIJI. In other countries, an interagency commission has been established to review and approve USIJI proposals. An example is the Russian Federation, where the head of the Russian Federal Service for Hydrometeorology and Environmental Monitoring chairs an interagency commission and reviews and approves projects on behalf of the commission and host government.

Reductions are additional

It is important that projects accepted into the USIJI programme do not simply constitute business as usual. The purpose of USIJI programme is to create new or 'additional' emissions reductions, not to provide certification to projects that would have occurred anyway. Therefore, in order to be accepted into the USIJI programme, proposals should:

- demonstrate that emissions will be reduced from what they would have been in the absence of the project. *This constitutes emissions additionality.*
- document that financing is in addition to normal Official Development

Assistance and is being provided because of USIJI participation. *This constitutes financial additionality.*
- certify that the project was initiated as a result of, or in reasonable anticipation of, USIJI. *This constitutes programme additionality.*

In practice, this has been a particularly difficult criterion to apply. As noted above, three distinct manifestations of additionality have been identified. Further definition, and some examples of the application of these interpretations of additionality, are provided below.

Emissions additionality Proposals should identify specific measures to reduce or sequester greenhouse gas emissions. It must be shown that, as a result of the project, emissions will be reduced from what they otherwise would have been. To demonstrate this, proposals must present a 'reference case', showing a baseline emissions scenario without the project, and a 'project case', which shows emissions projections over the life of the project. The difference between the project case and the reference case represents the emissions reduction.

The reference case has two important uses:

- It provides a reference point for historical greenhouse gas emissions and a projection of future emissions.
- It is a 'starting point' against which any future emissions will be compared.

Careful consideration is given to whether the reference case projections are consistent with

- prevailing standards of environmental protection in the country involved;
- existing business practices within the particular sector of industry;
- trends and changes in these standards and practices.

The reference case should provide information and data not only on greenhouse gas emissions, but also on other related environmental non-greenhouse gas effects.

In developing the project case, proposals should show how the specific measures identified in the proposal will reduce or sequester greenhouse gas emissions above and beyond those referred to in the reference case. Proposers are also encouraged to consider, as appropriate, off-site effects such as:

- Activity shifting – moving processes within an operation.
- Outsourcing – purchasing services or commodities formerly produced internal to the project boundaries.
- Market effects – offset to achievements caused by residual demand.
- Life-cycle emissions reductions – upstream and downstream changes in process materials used.

For both the reference case and the project case, considerable importance is placed on documentation of all resources, methods, emission factors and assumptions. Enough information needs to be provided in a proposal for an independent third party to understand all the assumptions that are made and be able to reproduce the emissions estimates and project effects. Various methods have been employed in proposal submissions. No single approach is endorsed by USIJI, nor does the use of a particular method imply acceptance. Instead, emphasis is placed on the adequacy of documented approaches to allow for third party validation.

Financial additionality Project funding should be independent of, or in addition to, the FCCC financial instrument, multilateral development bank or US government ODA, or, in the case of US federal funds, be in excess of levels provided in 1993. Project developers should demonstrate that in developing their specific USIJI proposal, they were able to receive financing that they otherwise would not have received. USIJI wants to be certain that the financial aspects of the project have been adequately considered, and that simple repackaging of federally or multilaterally funded projects does not occur.

Programme additionality Proposals should not only demonstrate that the proposed technology or practice reduces emissions, but that the technology or practice would not have been introduced but for USIJI. For example, if a technology or practice proposed in the project is required by an already established law or anticipated regulation, then the greenhouse gas emission reductions would have occurred anyway. If, however, the proposal shows that the emission reductions exceed what is required by law or international agreement, then the reductions may be considered to be additional, but only by the amount estimated that exceeds the legal requirement. Another situation might be one where the project proposes to employ a 'new' technology or method. In such cases, the net reductions would be considered additional, but only to the extent that it can be shown that the new technology was introduced as a result of USIJI. In other words, to be accepted, a project proposal should be able to show that the emission project for which emissions reductions are being claimed would not have occurred except for USIJI, or in anticipation of a similar programme.

Proposals can meet this criteria by showing that projects were formulated specifically for the USIJI programme. Projects can document that proposals were developed in response to workshops, or other outreach efforts of USIJI. Normally, a minimum requirement would be that project planning began after the inception of USIJI. However, in some cases, such as the Doña Julia Hydroelectric Project in Costa Rica, the project was actually conceived several years prior to the announcement of USIJI, but languished for a number of reasons. In these cases, it should be shown that USIJI was instrumental in overcoming barriers that would otherwise have prevented the implementation of the project.

Reductions are verifiable

Both verification and monitoring are important to assure the international community that real, measurable reductions are taking place. Monitoring and verification plans are required in proposals to make the process of measuring emissions transparent. Proposals must contain at least preliminary monitoring and verification plans. Where the proposed plans are less than adequate, the Secretariat will seek a commitment to improve these.

The plans should include adequate provisions for tracking the greenhouse gas emissions reduced or sequestered as a result of the project. Project developers are required to use the results of the monitoring process to periodically update estimates of emissions reductions and carbon sequestration. The monitoring and verification plans should address activity shifting or other actions that may result in 'leakage' of emissions outside the project site. For instance, if a project is presumed to result in reduced logging in one area as a basis for greenhouse gas reductions, the monitoring and verification plans should be designed to assure that logging does not increase elsewhere to compensate for the lost supply.

Proposers must also agree to a future process which may include verification of emissions reductions by third party organizations.

Reductions will not be reversed

Proposals should provide adequate assurance that greenhouse gas emissions reduced or sequestered will not be lost or reversed over time. For instance, a project proposal may estimate that it will sequester 100 millon tonnes of carbon (MtC) over the project duration of 40 years. However, if during years 41 to 45, the 100 MtC are released and this represents the end of the project, those reductions are not permanent.

The problem of reversal of effects is of less concern in energy projects where emissions reductions are generally considered irreversible. For example, if an energy project reduces emissions of a particular source from 20 MtC per year to 15 MtC per year for a period of ten years, the project has achieved 50 MtC reductions. This is true even if the emissions level rises back to 20 MtC per year at the end of the tenth year.

Other environmental benefits

Proposals should identify any associated non-greenhouse gas environmental impacts/benefits. For instance, a hydroelectric project may displace electricity generated by fossil fuel combustion, and as a result reduce emissions of other air pollutants. However, the hydroelectric project may have negative environmental impact on fisheries, water quality, or biodiversity. In reforestation and afforestation projects, planting trees in pastures and abandoned agricultural fields might provide positive secondary environmental benefits by stabilizing soil, restoring soil organic matter, and promoting the establishment of an understorey of native species brought in as seed by roosting birds. However, questions arise about the impact on biological diversity of planting when non-native, exotic tree species are used. The Evaluation Panel needs to be able to weigh the benefit of potential GHG emission reductions or carbon sequestration relative to any other positive or negative environmental impacts that the project might produce.

Annual reports

Participants must agree to provide annual reports to the Evaluation Panel on the emissions reduced or carbon sequestered, and on the share of such emissions attributed to each of the participants, domestic and foreign, pursuant to the terms of voluntary agreements among project participants.

Other considerations

In determining whether to include projects under the USIJI, the Evaluation Panel also considers several other issues:

(1) Leakage. This is the potential for the project to lead to changes in greenhouse gas emissions outside the boundaries of the project.
(2) Potential positive and negative effects of the project apart from its effect on greenhouse gas emissions. These include local employment and health impacts.
(3) Whether US participants are emitting greenhouse gases within the United States and, if so, whether they are taking measures to reduce or sequester these emissions.
(4) Whether efforts are under way within the host country to ratify or accede to the FCCC, to develop a national inventory and/or baseline of greenhouse gas emissions and sinks, and take measures to reduce its emissions and enhance its sinks and reservoirs of greenhouse gases.

Proposal solicitation process

Following the publication of the final ground rules and criteria in the Federal Register of 1 June 1994, the USIJI Secretariat developed a set of guidelines for preparation of proposals and announced that proposals under the first round of USIJI would be accepted until 4 November 1994. Following evaluation in accordance with the process described below, those projects that were determined by the Evaluation Panel to comply with the USIJI criteria were announced on 3 February 1995. A second round of proposals was solicited in May 1995, with a due date of 28 July 1995. Round 2 projects accepted in the USIJI programme were announced on 19 December 1995. The USIJI's Secretariat recently streamlined review process considers proposals three times each year.

Evaluation process

Proposals are reviewed not only from a purely technical standpoint, but also for the appropriateness of a particular project for the host country. Although this is not a criterion, the Evaluation Panel also examines the likelihood that projects will receive funding, as the goal is not just to approve good proposals, but to establish projects.

Using input from a team of technical reviewers, the USIJI Secretariat prepares a series of recommendations, or decision memoranda, for the Evaluation Panel. After careful consideration, the Evaluation Panel places proposals in one of three categories: accepted, placed in development, and not accepted.

Outreach

The USIJI programme performs a number of outreach activities, which are designed both to provide technical support and to identify project opportunities and partners. They are also mechanisms to relay general background information and programme status. The outreach effort is accomplished through bilateral and multilateral agreements, workshops, and print and electronic media. These activities are summarized below.

Bilateral and multilateral agreements on Joint Implementation

The US government has entered into bilateral and multilateral agreements with countries in various regions of the world in order to facilitate cooperation on joint implementation agreements. These 'Statements of Intent for Sustainable Development Cooperation and Joint Implementation of Measures to Reduce Emissions of Greenhouse Gases' (SOI) are designed to provide a framework for governments to cooperate on promoting private-sector investments in projects which fuel economic growth and benefit the environment. Key provisions in the SOIs include:

- Designation of a government contact on joint implementation with responsibility for creating programme criteria, and identifying, supporting and evaluating potential joint implementation projects.
- Information exchange on methodologies and mechanisms to establish procedures for monitoring and external verification of greenhouse gas emissions.
- Outreach and promotion of joint implementation and other sustainable development.
- Support of the international pilot phase at international fora.

As of 15 April 1996 bilateral SOIs had been signed with Bolivia, Chile, Costa Rica and Pakistan. A multilateral SOI was also signed between the US and Belize, Costa Rica, El Salvador, Guatemala, Honduras, Nicaragua and Panama. Most recently the Costa Rican and US governments signed an Annex to their bilateral agreement to facilitate a cooperative assessment of baselines and certifiable and transferable GHG offsets.

Information services

Fax-on-demand service Many documents pertaining to the USIJI process are available for delivery by facsimile. The USIJI Secretariat provides an automated fax-on-demand service.[1] Categories of documents available include:

- General background documents (e.g., a description of USIJI, the text of the final USIJI ground rules, submission procedures).

---

1   At (+1) 202-260-8677.This feature may not be accessible by all countries. Callers without access to a touch-tone phone may contact the Secretariat at (+1) 202-586-3467.

- Framework Convention on Climate Change (e.g., Decision 5 of the Conference of the Parties).
- Documentation from conferences and workshops (e.g., conference agendas, text of sections from conference notebooks).
- Results of USIJI submissions (e.g., list of projects, participants, and contacts for projects accepted).

Callers may select the documents through a menu-driven query process, or if they have a hard-copy menu (included with each fax delivery), they may skip directly to ordering. Up to three documents may be ordered with one phone call.

Newsletter The USIJI Secretariat publishes a periodic newsletter entitled *International Partnerships Report*, intended to provide updates on cooperative efforts to reduce greenhouse gas emissions. Issues have contained articles on the origins of USIJI, the signing of bilateral JI-related agreements, domestic and international USIJI workshops, the Evaluation Panel's selections from the first and second rounds of submissions, and the JI activities of other countries.

JI Online Under the sponsorship of the USIJI Secretariat, the Edison Electric Institute (EEI) International Utility Efficiency Partnerships Program (IUEPP) administers JI Online; accessible through the World Wide Web at http://www.ji.org. This computer bulletin board system is intended to enhance communication among people, institutions, and agencies working in JI, energy efficiency, and greenhouse gas mitigation projects around the world. JI Online provides a database of JI-related information from both government and private sources. Core libraries contain Secretariat publications, including project proposal guidelines, submission procedures, and contact information. Special areas can be set aside for materials from non-governmental organizations that wish to post information to the system. JI Online invites posting of any materials that might be of value to other people working on JI-related activities. Also included is a listing of potential USIJI projects which interested parties can review.

Accepted proposals

Fifteen projects from six countries have been accepted into USIJI, and are summarized in Table 11.1.

Future goals

The USIJI programme has several overall future goals that involve expanding the number of projects, as well as the depth and scope of experience, and providing technical support to project proposals at various stages of development. Individual goals include the following:

- Continue to test and develop criteria which demonstrate effectiveness in selecting projects that will produce real, measurable results.
- Increase the number of accepted projects and expand to new sectors and geographic regions.
- Implement a technical assistance programme for affiliated projects.
- Sponsor technical assistance workshops on emissions accounting, monitoring and verification, and financing of USIJI projects.
- Issue a technical handbook for project developers and a technical guidance document.
- Expand the USIJI public recognition programme.
- Assist countries in developing their national joint implementation programmes.
- Assist participants in obtaining project financing.
- Contribute to the better understanding of JI through analysing USIJI projects.
- Work with existing USIJI projects to confirm that estimates of GHG emission reductions, avoidance and sequestration are consistent and credible.
- Review, augment and develop, as necessary, monitoring protocols for existing USIJI projects.
- Develop a verification process or processes for application to USIJI projects.

Table 11.1 Projects accepted into the US Initiative on Joint Implementation as of 1 December 1996

Project title and description	Sector[a]

Belize

Rio Bravo Carbon Sequestration Pilot Project

This project combines (a) the purchase of a parcel of endangered forestland to expand existing protected areas, and (b) the development of a sustainable forestry management programme that would increase sequestration for a portion of the Rio Bravo Conservation Management Area (RBCMA) that includes the purchased parcel. A later phase will expand beyond the boundaries of the RBCMA — *Land use and forestry*

Costa Rica

Aeroenergía S.A. Wind Facility

A 6.4-megawatt private power wind facility, using 16 latest generation wind turbines, is expected to generate 27 GWh/year. This electricity will displace electricity currently generated by burning fossil fuels. Participants, which include a US system integrator and trading company and a Danish company that specializes in wind turbine technologies, currently anticipate financing the project with debt and equity — *Energy*

Biodiversifix Forest Restoration Project

This project will regenerate tropical wet and dry forest, expand the existing adjacent preserve, and implement sustainable forest manage–ment plans. Sustainable uses may include low-impact, controlled eco-tourism and minor regeneration of fine hardwoods. Income generated from carbon sequestration offsets and non-damaging biodiversity harvests will be used to cover project costs, as well as increase the endowment for future management projects — *Land use and forestry*

CARFIX: Project to Stabilize Existing Forest and Expand Forest Cover

Aims to stabilize existing natural forest within national park in central Costa Rica and provide additional forest cover in surrounding buffer zone, by instituting sustainable forest management, natural regeneration and reforestation. Funding for buffer zone activities will support outreach activities and annual payments to private landowners for progressive forest management, as well as purchasing private holdings in the park. — *Land use and forestry*

---

[a] See the IPCC Guidelines for National Communication for sector definition.
[b] The start date refers to the date the project will begin reducing GHGs.

Participants	Start date[b]	Project duration[c] (years)	Cumulative GHG emission reductions[d](tC)
US: The Nature Conservancy ● Wisconsin Electric Power Company ● Detroit Edison ● Pacificorp ● Cinergy. **Host:** Programme for Belize	1995	40	1,300,000
US: Power Systems, Inc. ● Bluefields, International ● Energy Works **Host:** Aeroenergía, SA	1997	20	9,800
US: The Nature Conservancy **Host:** Guanacaste Conservation Area ● National System of Conservation Areas ● National Institute of Biodiversity	1996	50	5,040,000
US: Wachovia Timberland Investment Management **Host:** Foundation for the Development of the Central Volcanic Mountain Range ● MINAE	1996	25	5,939,000

[c] Project duration refers to the estimated functional lifetime of the project, not necessarily the period over which GHG reductions are estimated to occur.

[d] Million tonnes of carbon equivalent. Reduction estimates are by USIJI project developers. The USIJI programme does not accept these estimates *per se*, but will be monitoring and verifying emissions reductions as they are attained.

Project title and description	Sector

Doña Julia Hydroelectric Project

A 16MW hydroelectric plant in northern Costa Rica, generating 90 GWh/year. This power will displace electricity that would have been generated from fossil fuels. Because of its small scale it will have less impact on the environment than a large dam. — Energy

ECOLAND: Esquinas National Park

Forestland inside a national park, which is under threat of conversion to agricultural use, will be purchased for management as a permanently protected tropical forest, and thereby preserve the carbon sequestration capacity. The project also will have other environmental benefits, including preservation of the region's biodiversity and habitats for endangered species, reduction of soil erosion, and maintenance of water quality. — Land use and forestry

Klinki Forestry Project

The proposed project would promote planting of Klinki pine intermixed with native tree species on private farms. Klinki pine is a very fast-growing, high-quality tree species native to Papua New Guinea. The trees will be propagated in a forest nursery of the Costa Rican partner. Long-term carbon storage would occur in trees (Klinki is long-lived) and in treated utility poles and other durable forest products. — Land use and forestry

Plantas Eólicas S.A. Wind Facility

A 20MW private wind power plant is scheduled to begin operation in spring 1996. The electricity generated will displace electricity currently generated by the burning of fossil fuel. — Energy

Tierras Morenas Windfarm

A 20MW wind power plant, displacing electricity that would have been generated from fossil fuels. The facility will use the world's first direct-drive wind turbines, which automatically adjust to changes in wind velocity for maximum efficiency. The project will be financed through private sources of capital, with a combination of equity and debt. — Energy

Czech Republic

City of Decin: Fuel Switching for District Heating System

Aims to reduce greenhouse gas and other emissions through a combination of (a) fuel-switching by a district heating plant from brown coal to natural gas, (b) cogeneration, and (c) distribution system improvements, to enhance the system's energy efficiency. The City of Decin, one of the most polluted cities in Northern Bohemia, also anticipates substantial health benefits from accompanying reductions in sulphur dioxide and particulate emissions. — Energy

Participants	Start date	Project duration	Cumulative Reductions
US: New World Power Corporation **Host:** MINAE ● Compañía Hidroeléctrica Doña Julia	1996	15	57,400
US: Tenaska, Inc. ● Trexler and Associates, Inc. ● National Fish and Wildlife Foundation **Host:** COMBOS Foundation ● MINAE ● Council of the Osa Conservation Area	1996	15	345,500
US: Reforest the Tropics, Inc. **Host:** Cantonal Agricultural Centre of Turrialba	1997	40	1,968,000
US: Merrill International, Ltd. ● Charter Oak Energy, Inc. ● Northeast Utilities ● KENETECH Wind power, Inc. **Host:** Plantas Eólicas S.A	1996	15	71,800
US: New World Power Corporation **Host:** MINAE ● Energía del Nuevo Mundo S.A ● Molinos de Viento del Arenal S.A	1997	15	51,000
US: Center for Clean Air Policy ● Wisconsin Electric Power Company ● Commonwealth Edison Company ● NIPSCO Development Company, Inc. **Host:** City of Decin	1996	25	165,600

Project title and description	Sector

Honduras

Solar-based Rural Electrification Project

The project will (a) replace kerosene lamps with solar-powered electric lights in rural regions that do not have electricity service, and (b) replace grid-based battery charging with stand-alone photovoltaic modules – in both cases reducing fossil-fuel emissions while generating. The project is based on a model developed by the US participant, a non-profit organization, which has already successfully field-tested the model in the region.

Energy

Bio-Gen Biomass Power Generation Project

This will develop a 15MW biomass waste-to-energy plant, funded through equity and debt. Located near a region with a substantial forest products processing industry, the plant will use as its fuel wood wastes currently disposed of through uncontrolled burning or dumping. It represents the first of three planned phases; when complete, the facility will have a 45MW capacity, the electricity being sold to the main Honduran utility.

Energy

Nicaragua

El Hoyo – Monte Galan Geothermal Project

This two-phase project will develop a privately owned and operated geothermal power plant northwest of Managua, with a 50MW facility to be on-line in 1999, upgraded to 105MW in 2001. The facility will use flashed steam technology, with hot water brought from the reservoir by deep wells. The host country participant is a Nicaraguan business developer.

Energy

Russian Federation

RUSAFOR: Saratov Afforestation Project

This project evaluates the biological, operational and institutional opportunities to manage a Russian forest plantation as a carbon sink in the context of joint implementation. It has planted 450 hectares of native tree species in degraded steppe grasslands that do not naturally generate forest, and reforested 50 hectares of burned pine forest. Benefits include reduced soil erosion, enhanced soil nutrient content, and local employment.

Land use and forestry

RUSAGAS: Fugitive Gas Capture Project

The project will make improvements to the natural gas distribution system and evaluate the technological, operational, and institutional opportunities to reduce methane emissions in Russia's natural gas production and transmission system. One of the US participants is a company with more than 25 years' experience of reducing fugitive natural gas leaks from pipeline valves.

Energy

Total cumulative GHG emission reduction (tC)

Participants	Start date	Project duration	Cumulative Reductions
US: Enersol Associates, Inc. **Host:** COMARCA ● AHDEJUMUR ● AHDE	1996	20	4,700
US: Nations Energy Corporation ● International Utility Efficiency Partnership ● Add-On Energy I **Host:** Biomasa-Generación	1998	20	647,400
US: Trans-Pacific Geothermal Corporation **Host:** C and R, Incorporated	1999	35	5,391,000
US: Oregon State University ● US Environmental Protection Agency (EPA) **Host:** Saratov Forest Management District, Russian Federal Forest Service ● International Forestry Institute (Moscow and Volga Regional Branches)	1995	60	35,000
US: Oregon State University ● US EPA ● Sealweld Corporation ● Sustainable Development Technology Corporation **Host:** GAZPROM ● Centre for Energy Efficiency	1997	25	8,182,000
			29,208,200

Chapter 12

Joint Implementation: an overview of an industry initiative and experience

Jan-Olaf Willums

Abstract

Joint Implementation as a political model had a difficult start, being criticized by developing countries as unfair and permitting industrialized countries to avoid their commitments made under the Framework Convention on Climate Change in Rio. A new approach is needed that changes the conceptual political model of JI and makes the technology transfer aspect the dominant driver. The policy negotiations can then focus on asset maximization and technology transfer benefits for the host country rather than solving the dilemma of equitable burden-sharing.

This chapter suggests such an approach, which could pave the way for a trading concept of carbon offsets. It outlines how the World Business Council for Sustainable Development (WBCSD) has approached this challenge through the International Business Action on Climate Change (IBACC) Initiative *Joint Implementation: An Overview of an Industry Initiative and Experience.*

Joint Implementation as global resource optimization: a business perspective

The United Nations Conference on Environment and Development introduced a political acceptance of efficiency and optimality, which led to an understanding of cost-effectiveness when looking for technological solutions. This was referred to repeatedly in the conference's main document, Agenda 21, but was also exemplified in the Framework Convention on Climate Change, which established the principles of returning to 1990 greenhouse gas emission levels by the year 2000 in the industrialized world.[1] The convention opened the possibility, although only vaguely for-

---

[1] Annex I countries include 24 OECD countries except Mexico, as well as 12 countries from Central and Eastern Europe with 'economies in transition'.

mulated, of sharing that commitment among countries.[2] The dilemma in the international negotiations since then has largely been rooted in the different interpretations applied to this concept by countries in the North and the South.

The idea behind joint implementation had already been put forward by various stakeholders, including the International Chamber of Commerce at the 1990 Bergen Conference 'Action for a Common Future'.[3] The proposal was to allow investment to reduce climate gases outside OECD countries to be credited to some unilateral commitments at home, and a first experiment of that concept was explored by the International Environmental Bureau between the energy industry in the UK and the Soviet Union in 1991.[4]

From the business point of view, the basic concept of JI, as understood by the parties drafting the Framework Convention on Climate Change, had appeal because as it reflected the business tradition of searching for efficiency gains wherever possible. Industry had experienced repeatedly that the cost of reducing the emission of greenhouse gases was lower in countries at a lower level of technological development: allowing commitments to be met by paying for the abatement abroad would clearly reduce the cost to meet legal commitments under the Climate Convention. At the same time, while the country carrying out the emission reduction might gain from local environmental improvement without incurring the costs, JI would encourage a flow of technology, which is a key element in most international business transactions.

The difficulties and pitfalls of JI

One problem, as seen from the business sector, was that central and eastern Europe and developing countries had not committed themselves to any reduction targets, and would thus have *a priori* no political or economic

---

[2]   Art 4.2. of the FCCC: 'The developed country parties ... may implement such policies and measures jointly with other Parties ... The Conference of the Parties, at its first session, shall also take decisions regarding criteria for joint implementation.'

[3]   Jan-Olaf Willums (ed.), *The Greening of Industry*, ICC Publications, Paris, 1990.

[4]   IEB Sustainable Energy Efficiency Programme 1991, Geneva.

incentives to reduce their GHG emissions. But these same countries have comparatively low GHG emissions reduction costs, and several corporations there felt that their economy would benefit from investments and technology transfer by the OECD countries which do have such commitments, and where reduction costs were much higher.

Although a global JI regime, which would make it possible for JI projects to be financed by the OECD countries and implemented by non-OECD countries, would indirectly mean that industry would encourage commitments towards binding targets and timetables –something that was not shared by all industry sectors – on balance a simple JI model would hold an important potential for becoming an efficient instrument for global GHG emissions reduction.[5]

The hope of rapidly reaching such a simple model has, however, faded since the Rio conference, and many industrial players have been losing interest in JI. Tom Heller of Stanford University has shown why the political economy of climate change has proved so difficult. He speculates on the prospects for escaping from the current deadlock over commitments through the multilateral diplomatic process, and has suggested reconsidering the concept of JI from a less orthodox perspective.[6] He has argued that the FCCC may not be an appropriate venue for solving the problems of a climate change regime, which mandates broad commitments by both developing and developed nations.

However, the FCCC is the only framework with a global commitment and therefore offers a chance of reaching a long-term global solution. Thus, JI remains, despite (or, as Heller points out, also because of) the political difficulties, the 'only game in play and – if reconceived – may provide an escape path from the deadlock over commitments in the multilateral negotiations'.

---

[5] Rolf Selrod, Lasse Ringius, Asbjørn Torvanger: *Joint Implementation – a promising mechanism for all countries?*, CICERO Policy Note, 1995:1, Oslo.

[6] Thomas C. Heller, 'Environmental Realpolitik: Joint Implementation and Climate Change', *Indiana Journal of Global Legal Studies*, Vol. 3, Issue 2, Spring 1996.

The requirements for making JI a viable concept

What, then, would be a reconceived notion of JI? Three key elements must be included: a political framework shift moving from a concept of *taking* to a concept of *ownership*, a focus on long-term technology cooperation, and a chance to introduce non-bureaucratic models of trading.

Moving from a transaction-based to an asset-based understanding of JI

The political image of Joint Implementation is that the rich countries move their problem to the poor countries and take away a benefit, i.e., an image of dumping a liability on your neighbour's doorstep. Let us exchange – just for the sake of argument – that image for one we are accustomed to from the resource sector, such as oil exploration. It is today generally accepted (although it was obviously not so in the early part of our century) that the ownership of a natural resource such as petroleum belongs to the sovereign state. Indonesia, say, is clearly the owner of its oil resources 'in the ground' – i.e. expected or believed to be there, but not yet explored or developed.

In order to 'mine' or explore this resource, a government may wish to do the seismic or geological survey itself, or hire others to do it. It may decide to drill itself, or ask others to drill and produce under a licensing agreement. Even then the oil produced is owned by the state, which can decide to either sell it (or keep it in the ground), or make an agreement with the explorer to share it in return for the technology, markets or capital provided by the exploration company.

Why can we not regard carbon offsets as a mineable resource that is clearly owned by the host government? If the government wishes, it can decide to enter into a development agreement in which a foreign entity provides the technology or capital that allows this carbon offset resource to be developed. The government may decide to pay for that service (or have the Global Environment Fund pay for part of it) and then sell the product to someone who wants to buy it – or keep it in the bank for later, if the price is not right. It may, on the other hand, decide to make a 'concession' deal where part of the resulting resource (i.e., a share of the carbon offset) is given to the technology provider in exchange for access to the needed technology.

Under such a conceptual policy model, there is no doubt about the host country's ownership and sovereign right to the resource. It can decide to use this resource as a negotiating tool (as many have done in exchanging oil rights for technology access), or decide to exchange it for more valuable assets (as Costa Rica has done in its recent negotiations of carbon offsets), or simply keep it in a bank. Or it may decide to keep part of the resource in the ground for later (as Norway once considered for its oil reserves).

If governments accept this model of carbon offset ownership, it may resolve the issues of additionality, as traditional development assistance is not involved. Also, the issue of whether a country providing technology has commitments or not is not up for debate – as long as a 'buyer' feels there is a need to secure carbon offsets.

Realizing the technology priming effect

The other conceptual model that has to change is the understanding of the value in having access to technology that goes far beyond what can be 'bought' in a technology licence deal. We need to shift from technology transfer to technology partnerships, and JI can have a strong priming effect of moving that kind of technology through partnership arrangements to the host country.

The World Business Council for Sustainable Development, a coalition of 125 corporations created to enhance the role of leading business in sustainable development, had already in its initial work on 'Changing Course' defined a new approach to technology cooperation that went well beyond technology transfer.[7]

What do we mean by a Technology Partnership? It is a relationship between business entities and technology organizations to develop the ability to use, modify, maintain, and constantly upgrade knowledge about an industrial product or process for the economic benefit of all parties involved. Technology Partnership is thus *more* than a commercial trans-

---

[7]   Stephan Schmidheiny, *Changing Course*, MIT Press/Business Council for Sustainable Development, 1992.

action; it is built on the joint wish for a long-term relationship between partners. It therefore ensures that environmental considerations are integrated into all investment and technology decisions, and includes the building of capacity for technological management, development and innovation.

Technology transfer in the past used to be equated with government-to-government programmes. Many of these transfers were clearly not sustainable. Although efforts were being made by various UN bodies and governments to improve government-to-government programmes, the UN Commission for Sustainable Development urged industry to become more directly engaged as trends were changing:

- technologies flowed more freely across borders, fuelled by electronic data communication;
- traditional aid budgets were shrinking, and international flows of private capital accelerating. Foreign direct investment from OECD countries is now three times greater than Official Development Assistance.

Environmentally sound technology, as it would have to be considered in a JI context, is not just pollution control, but technology insight to 'consume fewer resources and generate less pollution than the existing production system which it replaces. The development of environmentally sound technology is therefore a continuous search for new, cleaner and more efficient production processes.'[8]

Therefore technology cooperation can play a vital part in the implementation of international environmental agreements, including the Montreal Protocol, the Conventions on Biodiversity and especially the Framework Convention on Climate Change. We have to focus the JI debate on this notion of technology cooperation and experience sharing, and strengthen the real benefits of gaining access to such technology partners through JI.

---

[8]   Jan-Olaf Willums, Ulrich Goluke (eds.), *WICEM II: Proceedings of the Second World Conference on Environmental Management*, ICC, Paris, 1991.

Implementing a simple trading concept

The third requirement for a cost-effective climate change policy with strong support from business is a simple, unbureaucratic trading concept for carbon offsets. Any commercial development requires a functioning market, and once trading of carbon offsets is allowed to develop, the market forces will determine the real value of engaging in JI projects.

One condition is that transaction costs must be kept low. There are several interesting approaches to reducing transaction costs, which may well be evaluated when designing carbon credit trading. The Confederation of Norwegian Business and Industry has recently suggested that today's tax on sulphur in oil products be replaced by tradeable SO_2 permits within a national maximum emission level fixed at Norway's national target under the Oslo Protocol.

Under a proposed voucher scheme, individual companies would be permitted to buy or will be allotted tradeable emission permits in the form of vouchers. The total permitted emission would equal the national target. Used vouchers would have to be handed back, and unused vouchers may be saved or sold to other companies, but not used in advance. By controlling the number of vouchers, the government would be able to achieve a 30 per cent reduction of SO_2 emissions compared to estimates of the effectiveness of the present system. In addition, the cost to industry would be reduced by 75 per cent.

Other creative trading schemes and concepts that could define a market value and encourage JI have been suggested by other institutions and policy groups, as well as by UN bodies, such as UNCTAD, and may contain elements that are applicable and useful for a radical rethinking of tradeable permits.

The International Business Action on Climate Change

One problem with the projects suggested under the AIJ pilot phase is that most of them are suggested by companies or consultants in the North. Most of the initial AIJ projects have been initiated by energy companies in North America or northern Europe, under the sponsorship or encouragement of national governments. In early 1996, therefore, the World

Business Council for Sustainable Development launched an initiative that would encourage business in *developing countries* to propose and initiate partnership projects with counterparts in the North, the International Business Action on Climate Change, focused on generating project proposals by developing countries and central and eastern Europe.

The initiative generated an interest in seeing JI as a driver for the transfer and dissemination of climate-friendly technologies, even if the prime purpose of IBACC was to help companies and institutions in emerging economies to initiate business-to-business partnerships and implement projects which have a positive climate effect and which may thus be eligible for support under the concept of 'Joint Implementation' or 'Climate Technology Cooperation'.

The WBCSD wanted especially to encourage developing-country enterprises to become more active in promoting and seeking commercially sound projects, and thereby to convince their governments that the JI concept is a viable and useful approach also for developing countries and countries in transition.

On behalf of the WBCSD, TransAlta of Canada drafted a WBCSD 'Call for Projects' and launched it on the World Wide Web in February 1996. In parallel the WBCSD arranged a series of capacity-building workshops and briefings to explain the concept of JI and help companies and institutions to identify and prepare better project proposals in El Salvador, Bangkok, Singapore (jointly with the Research Institute for Environmental Technology), and Hong Kong, and later in Tallinn, Estonia (co-sponsored by the Stockholm Environment Institute) and Prague (in cooperation with the Czech BCSD and the Dutch JI Foundation).

By June the request for proposals had generated over 100 project ideas, of which 80 project proposals were being submitted in the required format. They were then screened by a team in Canada and evaluated in more detail. Projects were submitted from Latin America (Nicaragua, Bolivia, El Salvador, Brazil, Colombia, Guatemala, Mexico, Costa Rica), Central and Eastern Europe (Ukraine, Russia, Armenia, Poland, Czech Republic, Estonia), Africa (Cameroon, South Africa) and Asia (Indonesia, Philippines, India, Nepal, Sri Lanka, Thailand).

The projects suggested for the IBACC initiative covered energy efficiency, renewable energy projects, sequestration, methane reduction, waste management, power conversion/fuel switching, composting, and transport efficiencies. The total carbon value of the proposed projects was 115 million tons of carbon equivalent to 2300MW produced over 20 years. The size of potential investment in the 35 most realistic projects represents an investment of US$ 293 million, with an initial rate of return from these projects ranging from 15 per cent to 40 per cent.

As part of a cooperation arrangement, the World Bank Group then evaluated the most promising project proposals in detail. The next step in IBACC will be a detailed project proposal evaluation and a selection of a number of key projects that can demonstrate the benefits, but also demonstrate the policy framework that is required.

Conclusion

We believe that the concept of JI is still a highly valuable policy direction to be explored and improved. That requires a rethinking of the basic concept of JI as an *asset-generating value concept* for the host country rather than a transaction process to avoid commitments under the FCCC in the North. The technology transfer aspect of JI will thereby become the dominant driver, and policy negotiations should focus on optimizing the multiplier effect of technology flows, and encouraging JI developments by providing the framework for an efficient market-place for carbon offsets.

The WBCSD regards a rational and effective model for *measuring and crediting* as a key element for the success of JI. A proposal was presented at the Prague workshop, and is now being developed further towards some dialogue and concrete recommendations in cooperation with the World Bank and the Academy of the Environment. The goal is to move the crediting discussions forward as far as possible by COP-3, thereby increasing the chance that JI may become the tool to enable a global approach to the climate challenge.

Chapter 13

Japan's programmes for Activities Implemented Jointly during the pilot phase

Katsunori Suzuki and Katsuo Seiki

Introduction

The Conference of the Parties to the United Nations Framework Convention on Climate Change at its first session decided to establish a pilot phase for Activities Implemented Jointly, which is open to the participation of developing-country Parties on a voluntary basis.

Recognizing the importance of a global reduction in greenhouse gas emissions and the cost-effectiveness of Joint Implementation in controlling and lowering such emissions, it is important to study and promote AIJ, which could be considered an effective tool for transferring technological and financial resources, particularly through the private sector, as well as a trial phase of JI.

Progress on AIJ in Japan

The government of Japan has been active in promoting AIJ. It established the Japan Programme for AIJ under the Pilot Phase in November 1995, and in January 1996 issued the 'Manual for AIJ Project Applications, Guidelines for Approving AIJ Projects' and an AIJ project application form. The government started to receive applications for AIJ projects from April 1996 and in July 1996, on the basis of the applications, it identified eleven projects as potentially promising, including one combustion improvement project, two power-sector energy efficiency improvement projects, one solar project, one coke dry quenching process (CDQ) project, and six afforestation projects. The eleven projects are summarized in Table 13.1. Japan will assist in promoting them subject to the consent of host governments. Under the AIJ framework, the government of Japan seeks wide participation by the private sector, local governments and NGOs on a voluntary basis, with a view to providing other Parties with extensive experience in AIJ.

Table 13.1: Promising projects identified by ministries and agencies for AIJ

Project name (project proponent)	Category (target GHG)	Location (participants)	Competent agency
Combustion improvement project for small-sized coal boilers (Kitakyushu City)	combustion improvement (CO_2)	Dalian, China (Dalian City)	Environment Agency
Local electrification project using solar panels (Solar Net)	solar energy use (CO_2)	Salatiga, Indonesia (YAYASAN GENI)	Environment Agency
Heat efficiency restoring project of existing thermal power generation plants through improvement of operation (power companies)	energy-saving analysis & improvement in terms of operation & maintenance (CO_2)	Thailand (EGAT)	Ministry of International Trade and Industry (MITI)
Local electrification project (E7)	solar & hydraulic energy use (CO_2)	Indonesia (Science & Technology Application Agency)	MITI
Model project for coke dry quenching process for coke furnace (NEDO)	waste heat recycling (CO_2)	China (Metallurgic Division)	MITI
Afforestation in Sabah (Kokusai Ryokuka Suisin Center)	sink enhancement (CO_2)	Sabah, Malaysia (Sabah Forestry Public Association)	Ministry of Agriculture, Forestry & Fisheries (MAFF)
Afforestation (same as above)	same as above	West Nusa Tenggara, Indonesia (Ministry of Forestry)	MAFF
Afforestation using local tree species (Nissei Green Foundation)	same as above	Kenya (Kenya Forestry Research Institute)	MAFF
Afforestation for volcanic desolated area (same as above)	same as above	Bali, Indonesia (Ministry of Forestry)	MAFF
Experimental afforestation (Sumitomo Forestry Co. Ltd)	same as above	East Kalimantan, Indonesia (Ministry of Forestry etc.)	MAFF
Afforestation of desert rim area (Chikyu Ryokuka Center)	same as above	Inner Mongolia, China (Inner Mongolian local administration offices)	MAFF

International cooperation

JI is a concept provided for in Article 4. 2. (a) and (d) of the UNFCCC. However, AIJ was a new concept introduced by COP-1, which agreed to establish a pilot phase among Annex I Parties and, on a voluntary basis, with non-Annex I Parties that so requested. It is necessary to accumulate experience in AIJ projects under the pilot phase in order to generate an international consensus on the fundamentals of JI, as well as facilitating the transfer of environmentally sound technologies.

To facilitate the process, the Environment Agency of Japan and the International Environmental Technology Transfer Centre of the United Nations Environment Programme (UNEP/IETC) held a Regional Workshop on Transfer of Environmentally Sound Technologies and Activities Implemented Jointly, on 19–21 June 1996 in Osaka. The objectives of the workshop were to:

(1) disseminate information on activities regarding the transfer of environmentally sound technologies (ESTs);
(2) review progress of on-going international activities on the transfer of ESTs and identify shortcomings of such activities;
(3) exchange information on AIJ; and
(4) assess the viability of innovative approaches such as AIJ to further facilitate technology transfer, and explore effective means of optimizing the use of AIJ in meeting the objective of the UNFCCC.

The workshop was attended by experts from ten countries in the Asia-Pacific region, and representatives of six international organizations.[1] The Environment Agency also co-sponsored:

- the International Workshop on AIJ, 25–27 June 1996 in Jakarta, together with the government of Indonesia and the US Initiative on Joint Implementation (USIJI); and
- the Forum on 'New Partnerships to Reduce the Build-up of Greenhouse Gases', 29-30 October 1996 in San José, Costa Rica.

---

[1]   The chairperson's summary of the workshop is reproduced in the full proceedings of this conference, or can be obtained from the Japan Environment Agency.

The Japanese government will continue its efforts to promote AIJ Pilot Phase projects, and to provide information on the phase to the international community.

Framework of Japan's Programme for AIJ

The main points of the Japan Programme for AIJ under the Pilot Phase are as follows:

Objectives

(1) To accumulate experience in order to contribute to discussions on the formation of an international framework of JI to be implemented in the future;

(2) To establish a methodology for measuring, in a comprehensive manner, the net reduction or absorption of GHG emissions to be achieved by AIJ/JI;

(3) To formulate steps to encourage the private sector to participate in future JI projects.

Eligibility

Domestic participants:

(1) Japanese nationals and/or residents in Japan;

(2) Corporations, organizations, associations and other legal bodies in Japan;

(3) National and/or local governments; and

(4) Other entities recognized as being able to carry out an AIJ project.

Foreign participants (including, but not limited to, Annex I countries):

(1) National and local governments of the country Parties to the Convention;

(2) Nationals of, and/or residents in the above country Parties;

(3) Corporations, organizations, associations and other legal bodies in the above country Parties; and

(4) Others in the above country Parties, who are recognized as being able to carry out an AIJ project.

Implementation Mechanism

The Japanese government has established an Inter-Ministerial/Agency Coordination Committee for AIJ (IMACC) to facilitate communication among the ministries and agencies concerned and to ensure the above-mentioned processes. Competent ministries/agencies will evaluate an application for an AIJ project and approve it, if appropriate, as an AIJ project. The IMACC is structured as follows:

Members of IMACC
- Environment Agency (co-chair);
- Ministry of International Trade and Industry (co-chair);
- Ministry of Foreign Affairs; and
- Other relevant ministries and agencies.

Major tasks of IMACC
(1) To authorize the evaluation guidelines for AIJ projects;
(2) To undertake hearings on the progress of AIJ projects to be reported by the secretariat, and to prepare annual progress reports on AIJ projects;
(3) To encourage private entities to participate in AIJ;
(4) To coordinate issues relating to the implementation of the Japan Programme;
(5) To review and rearrange the present AIJ implementation mechanism, as appropriate.

IMACC Secretariat
The IMACC Secretariat consists of the Ministry of Foreign Affairs, Ministry of International Trade and Industry and the Environment Agency, and was set up to facilitate communication among ministries and agencies concerned.

Application procedure

Those who undertake AIJ projects shall submit an application for the AIJ project in accordance with the Manual for AIJ Project Applications to the competent ministry/agency, or to the IMACC Secretariat if the responsible ministry/agency is not identified by the project proponent.

Upon receipt of an application document, the competent ministry/agency shall send a copy to the IMACC Secretariat. If the IMACC Secretariat receives an application, it shall send the document to the competent ministry/agency.

Evaluation, approval and reporting to IMACC

The competent ministry/agency shall evaluate applications and approve them if appropriate, in accordance with the Evaluation Guidelines. The competent ministry/agency shall provide guidance/advice to the project proponents, if there are problems in implementing their project.

The competent ministry/agency shall report to the IMACC Secretariat on the evaluation and approval as well as the status of the implementation of the approved AIJ projects.

Evaluation criteria for AIJ projects. The competent ministry/agency which is to supervise a relevant AIJ project shall ensure that the proposed project satisfies the following requirements, in approving it as AIJ under the Japan Programme:

(1) Predicted results of GHG emissions (or absorption) shall be presented both with and without the proposed project, together with sufficient information on the prediction. Projected emissions within the project shall be less than would have occurred without the project, or absorption projected within the project shall be more than would have occurred without the project.

(2) Cumulative effects of GHG emission reduction resulting from the proposed project shall not be negative.

(3) Project implementation entities shall regularly review predictions and modify them as appropriate. They shall inform the competent ministry/agency of the modification.

(4) The proposed project shall be additional to the financial obligations of the Parties set out in Article 4.3 of the UNFCCC as well as to current official development assistance (ODA) flows.

(5) With the official consent of the host government, the proposed project shall be regarded as an AIJ project.

(6) The proposed project shall not cause a greater increase in GHG emissions in other areas than the expected reductions through the project.

(7) The environmental, economic and social impacts of the proposed project shall be adequately assessed. The results shall be reviewed by relevant authorities as appropriate.

Lessons learned

The Environment Agency of Japan has been supervising two potentially promising AIJ projects, one in China and one in Indonesia. On the basis of experiences and discussions at various international fora on this topic, the following lessons have been learned.

(1) Institutional arrangements for an AIJ framework in both donor and host nations are essential for identification and implementation of AIJ projects. The inter-agency coordination mechanism seems to be very effective for better communication and coordination among relevant parties.

(2) An endogenous capacity-building in host countries also seems to be crucial. An AIJ project may not be able to cover all the components for technology transfer and capacity building required for the proposed project. Cooperation and coordination on technology transfer with other appropriate projects seems to be very effective.

(3) There is some confusion on methodologies, especially for baseline calculations and monitoring. It is highly desirable to establish and disseminate common methodologies.

(4) Given the lack of any crediting of emission savings to the investor during the pilot phase, strong government initiatives and support seem to be necessary

(Katsunori Suzuki)

AIJ projects promoted by the Ministry of International Trade and Industry

Of the eleven projects authorized by the Japanese government, three were related to the Ministry of International Trade and Industry (MITI), i.e., the applicants were from the industrial sector:

(1) *Model project to install coke dry quenching equipment to coke furnaces in China*

This model project concerns coke furnaces at integrated steel works. To produce coke for injection into a furnace, coal is heated and dry distilled in a coke furnace. In the conventional process, the heated coke is cooled by water, thereby dispersing and wasting heat, with possible environmental effects such as air pollution etc. Currently, however, the dry quenching method, adopted at Japanese steel works, uses inert gas instead of water in the quenching and cooling process. Inert gas enables the reclamation of waste heat to generate high-temperature, high-pressure vapour which, in turn, can be utilized at steel works to achieve considerable energy savings.

The New Energy and Industrial Technology Development Organization (NEDO) is a major Japanese participant in this model project in China.

(2) *Local electrification project in Indonesia*

This will be a joint project between the world's major electric power companies from the United States, Canada, Germany, France, Italy, and Japan. From Japan, Tokyo Electric Power Company (TEPCO) and Kansai Electric Power Company (KEPCO) will participate. The plan is to proceed with the rural area electrification projects in Indonesia – where the rate was 44% in 1991 – through the installation of power-generating equipment such as home-use solar batteries (around 50 kW capacity), mini-hydraulic-power plants (about 200 kW), and hybrid system (solar batteries with back-up diesel generators). The project will be implemented in cooperation with PLN (Indonesian Electric Power Utilities) to promote efficient power generation and utilization in rural areas, while striving to conserve the global environment.

(3) *Project to restore thermal efficiency of existing power plants in Thailand through operational improvement*

This project is aimed at improving the thermal efficiency in power generation plants in cooperation with EGAT. It is based on various energy-saving measures for the operation and maintenance of power plants obtained from the existing thermal power plants of Japanese electric power companies. The major participants from Japan are KEPCO, Chubu Electric Power Company (CEPCO), and the Electric Power Development Company (EPDC).

Apart from these three projects authorized by MITI, there are six afforestation and reforestation projects authorized by the Ministry of Agriculture, Forestry and Fisheries, and two authorized by the Environment Agency. With the exception of the Kenyan afforestation project, all these other projects are in East Asian or Southeast Asian countries, whose vigorously growing economies are expected to generate further and faster increases in CO_2 emissions. Profound effects on regional and global emissions are anticipated from the implementation of these projects.

The AIJ Promoting Forum

Apart from the AIJ activities in which industries and the public sector are the main participants, the Japanese government has also established an AIJ Promoting Forum to involve the interested parties.

(1) *Purpose.* To provide an opportunity for local governments, industries and NGOs to exchange ideas and opinions, as well as periodically to provide information to actively promote the AIJ Japan Programme.

(2) *Participating members.* Mainly those parties mentioned above, including the members of the Inter-Ministerial Agency Coordination Committee for AIJ. The executive office of the Forum will be located at the Global and Human Environment Forum Institute and the Global Industrial and Social Progress Research Institute.

(3) *Meetings.* The 'Preparatory Meeting of the Forum for the Promotion of AIJ' will involve industries, local government and NGOs and the forum be officially formed following the authorization of projects.

(4) *Main activities:*

- (a) Providing information: hold explanatory meetings for the project applications. (Before the start of first-phase project applications, such meetings were held twice in Tokyo and Osaka. The executive office of the Forum hosted these meetings.)

  (b) Providing AIJ policies and monitoring information from each nation to major participating entities. (Currently AIJ-related information is offered through GISPRI's home page on the Internet.)

- Selection and administration of a logo: the executive office of the Forum has appointed a Logo Selection Committee which selects the logo for use in AIJ Japan Programmes. The Committee will be managed by the executive office of the Forum.

- Exchange of opinions/views between public and private sectors on issues related to the implementation of projects authorized under the AIJ Japan Programme.

- International workshops etc. It is planned to hold meetings among Asian nations to exchange views on AIJ.

(Katsuo Seiki)

Chapter 14

Emissions trading in the context of the climate change negotiations

Michael Grubb

This chapter explores the relationship of potential CO_2 trading and joint implementation to the evolution of international and national law and politics, as mediated through the climate convention processes. I will be looking first, briefly, at why emissions trading is attracting this kind of interest and at its relationship to the comprehensive approach, using multiple gases. Then I consider the question of which sectors we are talking about controlling in connection with tradeable permits, and what that implies; the politics of the process, particularly relating to the European Union rather than the United States; and, more broadly, the international perspective beyond the OECD in the relationship with joint implementation. Finally, I touch upon the role of the Kyoto protocol.

Why emissions trading?

Why are we finding ourselves in this debate? There is the classic economic argument that tradeable permits are an efficient way of implementing emission reductions, but then in principle, if the market is working well, so too are other economic instruments such as a tax. Dirk Forrister suggested that in the United States the energy tax proposals had the minor drawback that it helped to lose the Democrats' control of both Congress and the Senate (see Chapter 2). So the question is, why is there such a big political difference between instruments that many economists say are very similar?

There are some economic differences between caps and taxes noted in the economics literature, relating to uncertainty and whether one is trying to control the product or the cost incurred. This argument suggests that taxation might in fact be the more appropriate instrument for CO_2 control.

But the politics point the other way, for key reasons which have not appeared much in the literature. A trading instrument creates an asset.

Even if the permits are auctioned and industries have to buy permits to cover emissions, the financial transfer is the same. The company now has an asset on its books, so from the industrial perspective it is different. If permits are granted or given out initially, then the company has acquired an asset, and it looks totally different in terms of the industrial politics for that reason. Also, within countries governments have a negotiating flexibility over the relationship between auctioning permits – selling them – and giving some out. There is a lot of political flexibility for the governments, crudely, to buy off the most powerful parties.

Finally, there is an institutional appropriateness in the sense that energy or carbon tax instruments are perceived as a tax instrument. This means that by and large they became the responsibility of the finance ministries and treasuries, whose primary objective is not limiting CO_2 emissions. Frequently, indeed, we find a big institutional mess in governments about who is controlling the policy. Permits are more clearly an instrument that falls in the legitimate terrain of environment ministries. These are very big political differences which explain why we are now looking down this road.

Jørgen Henningsen made some interesting comments about some of the drawbacks and limitations of tradeable permits, particularly if we are still talking mainly about exploiting rather cheap emissions limitations.[1] In a situation where some people are saying this is a problem that will cost us the earth, and others do not even expect the Kyoto protocol to result in any measures that will cost anything, tradeable permits would at least help to reveal who is right. They would reveal prices, and potentially put limits on them.

Comprehensive approach
The simple comprehensive approach

The approach of including all the different greenhouse gases and sources as a basket is intellectually extremely appealing. It increases the flexibility and efficiency of the process overall. Before the Convention in 1992 the United States gave this considerable political support. The Europeans were very

---

[1]   For a short discussion of the contribution of Jørgen Henningsen see Chapter 1. See conference proceedings for full text.

hesitant, however, and this resulted in a compromise in the Convention. The phrase used is 'CO_2 and other greenhouse gases' – one of the many creative ambiguities in the Convention as to whether the CO_2 should be regarded separately.

Interestingly enough, more than four years later, during the AGBM negotiations in December 1996, in its submission to the protocol the US focused on CO_2 trading, whereas the European Union had dropped the words 'CO_2 and other' and referred only to 'greenhouse gases'. The UK submission stated that 'there is no longer any reason not to adopt a comprehensive approach'.

Despite its theoretical appeal, there is a big problem with the comprehensive approach, particularly in the context of legally binding targets and permits. If you intend to make something legally binding, still more to attach some kind of financial value to it, as would happen with trading, you need to be very sure about the monitoring capabilities – the costs of monitoring, accuracy, verifiability, and the basis on which you are comparing different gases.

There is a whole range of gases and sources, starting with carbon dioxide from fossil fuels and running through the different kinds of CO_2 sources, the halocarbons, methane from a whole range of sources. These differ enormously in terms of the ability with which they can be monitored. Among the methane sources we have bovine flatulence, which in greenhouse terms, I guess, is the biological equivalent of car exhaust, in the form of cow exhaust. Clearly no one can monitor methane from bovine flatulence to within the few per cent accuracy that would be required for a tradeable permit system. It is not credible to combine the two in that way, and that needs to be made quite clear.

Two lists approach

We can think of a more feasible approach, if it is worth the complexity. A protocol could establish a system which starts out with what we can do in terms of the monitored and verifiable sources in a first list, and which would also specify how the different sources were to be compared. It could then establish a second list of sources which we could not be controlled in that way but which would still be addressed in other ways. This is important. There needs to be pressure on the other sources and sinks, for example with non-legally binding targets as appropriate – because you cannot monitor a binding target – and perhaps tackled by specified policies and measures. If – and that is an important *if* – it is considered important enough, again the protocol could establish a process by which one would seek to move over time different sources from list two to list one, as and when it was institutionally feasible.[2]

US sulphur model

Experience

I now want to look at the US model extrapolated from the sulphur experience and applied to CO_2. Relatively little has been said about the fact that this model is limited to certain sectors – in the US, power plants above a certain size; in the EU one could conceive that a very similar sort of system would be confined to boilers that fall under the purview of the Large Combustion Plant Directive for Sulphur, namely above 50 megawatts thermal capacity. Of course, this is not relevant to the use of 'trading' for exchanging national emission quotas, but it is important for the use of tradeable emission permits as an instrument of domestic implementation.

In terms of climate change, this immediately raises the question of the non-participant sectors. Would such a system not introduce distortions of scale, if the small boilers profit, or of sectors – is it not unfair, the power generators would say, that other sectors are not properly controlled? That will lead to political repercussions. Some thinking, therefore, about the non-participant sectors is required. They may be addressed through subsectoral targets – what governments are expecting each sector to contribute – or specific policies and measures.

There is a perception that the US is pushing emission targets and trading, while the European Union is pushing policies and measures. The chances are that one needs to look at and combine both, because you need different kinds of instruments for different sectors, at least if one is looking primarily at carbonizing the US sulphur model.

---

[2] For a further elaboration of this, see M. Grubb, D. Victo, and C. Hope, 'Pragmatics in the Greenhouse', *Nature*, Vol. 354, 5 December 1991, pp. 348–50.

Obstacles to internationalizing the US model

We need to be very careful about what we are getting into. Rick Bradley covered some of this,[3] but going international changes the system a lot. Some countries have very different energy economic structures. For example, Switzerland, Sweden and France have essentially very close to zero carbon in their power sector. In Switzerland, only 13 per cent of carbon dioxide emissions comes from power generation or from heavy industry, so a carbon version of the US model is almost irrelevant. The Swiss do not gain any serious flexibility from it, and their concern is with other sectors.

Similarly there may be other kinds of problems or differences in energy economic structures – a few countries may have large electricity imports from non-participants in the scheme. These need some attention.

There are also countries with different legal structures. I have been told that in neither Canada nor Australia do the federal authorities have the legal powers to implement a tradeable system of the kind in which the US is involved, and other countries have very different legal and institutional relationships.

Perhaps more important than any of these issues, countries have very different cultures and institutions related to public policy. Clearly France and Japan have very different assumptions underlying the relationship between government and industry than, say, the United States. Indeed, much of continental Europe has seen a preference for taxation and negotiated agreements to grant tax exemption to some industrial sectors. Likewise countries differ in their monitoring and enforcement capabilities.

Personally I am very much in favour of US emissions trading, but I believe it will take different forms – maybe very different forms – in different countries. What the Kyoto protocol will not do is create a coherent international trading system; but it could establish a framework of national commitments within which different national trading systems can be developed and made internationally compatible.

---

[3] A short discussion of his view is given in Chapter 1. His paper can be found in the conference proceedings.

Alternatives to the US model

There are a number of options. Countries might have slightly different sectoral coverage of trading systems. They can try to apply the permits at different points in the production chain – for example, to the carbon content of the fuels coming out of refineries. I have heard suggestions by both Canada and the European Union that maybe they want to focus on the traders, the companies that deal in fossil fuels, rather than the companies that burn them. I have serious reservations about the feasibility of that approach, but I may be wrong and I know it is being considered.

Likewise geographical coverage; I mentioned the Australian federal problem, and I was most interested to learn that New South Wales is considering trading within the state. Governments themselves could hold and trade emission quotas, as the basis for enforcing national emission commitments. It is an idea which sends shivers down the spine of some US industrialists I know, but not all European countries or publics or industries have quite the same aversion to governments being involved in that way.

CO_2 trading in Europe

The situation in Europe is very tricky and may illuminate some of the issues. Why should Europe be interested in trading, despite all the caveats that Jørgen Henningsen raised?[4]

The most obvious reason is the complete and rather abject failure of most EU-level policies to date, the most obvious of which is the carbon tax. But the same applies to some extent to other policy proposals. They have tended to fall on the altar of EU politics and subsidiarity. I cannot resist mentioning here that about three years ago we had a meeting which involved a discussion between Ernst von Weizsäcker in Germany and myself on the question of policy instruments. People had been looking forward to a vigorous argument between the proponent of taxes and the proponent of permits, and they were very disappointed when we both agreed that at the international level quota trading made more sense, because you did not want to interfere in the national sovereignty of instruments that

---

[4] See Chapter 1.

countries used. At the national level, certainly in some sectors, taxation may well make more sense.

Having apparently found some consensus, Ernst and I decided to write something about this together. I wrote the first draft, on European policy, and faxed it to Ernst. A horrified reply came back; we had not realized that I had been thinking of the European Union as essentially a collection of states, whereas Ernst had been thinking of it as a unified state. That may tell you something about the nature of the European problem! Nevertheless, the upshot of that dialogue was a report which eventually emerged on emissions trading involving a number of European institutes.[5]

I think it is unlikely that the European Union will magically acquire powers to deal with issues such as taxation or energy policy, which did not even appear on the agenda of the Intergovernmental Conference. And yet, if we look at the national programmes, there are a lot of no-cost or low-cost measures, but there is a process of gradual exhaustion which will lead to more and more pressure to look at more potentially costly measures, in economies which are increasingly closely entwined. That trend is most dramatic in the electricity sector, where the June 1996 Liberalization Directive establishes a process towards opening up all the member states' electricity sectors to international competition in Europe. That will set limits on what member states can do without having instruments coordinated at European level.

Another reason why the EU needs to find a feasible approach to the coordination of instruments, I would suggest, is the international politics of the protocol negotiations. Clearly a number of countries outside Europe are watching very keenly to see the credibility of European commitments in this area. Indeed, I know one government represented at this conference has said it will block having a Union signatory of a protocol unless it is very clear how the Union is itself going to deliver what it is calling upon other countries to do. That requires the instruments that will be used at a pan-European level to have some credibility.

Finally I have already made reference to the legal and institutional basis of policy instruments. The European Union does not have legal compe-

---

[5] M. Grubb et al., *Implementing the European CO_2 Commitment: A Joint Policy Proposal*, RIIA, London, 2nd edition, February 1997.

tence in the fields of taxation or energy policy overall, but a tradeable permits system is arguably an emissions control/environmental policy which has a legal basis of competence, and which in its institutional design may not even require unanimity in the voting procedures. There are thus institutional reasons why Europe might profitably take a closer look at emissions trading.

All this may sound fine in theory, but will it happen? I would agree the prognosis is not very good. There are a few individuals in a few European governments who recognize essentially that the European Union does not yet have clear and credible proposals for implementing the commitments it is calling upon the rest of the industrialized world to agree to. The Commission itself is discussing this but is very much divided on questions of permits trading, and anyway could not drive such a big change. Politically the only way one could see it emerging would be if one country started to champion the idea.

The European trading system

On the Kyoto time-scale the need is not to design and create a full tradeable permit system, but to establish the conceptual and political basis that would enable the Union credibly to sign legally binding commitments and over subsequent years to start to implement them, including but not exclusively using trading programmes. If that were to occur, then it would probably start with the concept of intergovernmental quotas – exchange of national targets. This might become mixed in with permits which are diffused down into some sectors, power generation being perhaps the most obvious, following the process of electricity liberalization.

My guess is that from an early stage border relationships would be established, bringing Norway, as Berndt Bull mentioned in Chapter 9, into the European emissions bubble. There is a huge mutual interest in enabling that to happen, given Norway's domestic predicament combined with its role as a key gas – and increasingly electricity – exporter to the Union.

Politically then, having established some credibility in the process, the key is how Europe would address the implementation issues after the protocol, after signing on to a collective commitment. Interestingly, the time-scale is quite good if that occurs. A European CO_2 trading system will not be developed by a group of ministers sitting round a table and negotiating a fair allocation. It will involve messy, bloody questions of the politics of allocation within Europe, but at the end of 1997 Europe also enters a whole new cycle of budgetary negotiations – revision of the structural funds, and associated negotiations on enlargement.

If Europe does seek to implement its Kyoto commitment with intergovernmental quota trading, allocation will become part and parcel of that horse trading. It could become intimately linked with proposed changes in structural expenditures in the cohesion countries, for example. It seems to me that within that rather messy political process this is in practice how any allocation negotiations will proceed.

Beyond that basic process one would then have to start seeing the development of an implementation architecture, presumably involving European-level institutions for monitoring and carrying out the exchange, sectoral definition of exactly which sectors are involved, and ultimately leading to a carbon currency. Although the prospects on both fronts are problematic, one might conceivably see a common carbon currency before we get to the common European currency!

Joint Implementation

Let me come to the question of Joint Implementation among Annex I countries. I will not say anything about Activities Implemented Jointly outside Annex I, other than that attempts to reintroduce that question in terms of official intergovernmental trading or offsets for the Kyoto protocol may risk undermining the hard-won political basis of the Berlin Mandate and the decisions on the activities implemented jointly. JI is currently an issue within Annex I, and one should proceed first on that basis.

In central and eastern Europe JI projects are beginning to happen. This has political legitimacy, and will be tied in some sense to the generation of credits, as was discussed earlier, which might become formalized into sectoral allocations leading to permit trading. I do not see this as being a very neat or uniform process; the dissemination of these instruments will occur at varying rates.

In Europe, this development will also be rather intimately related to energy trade, particularly gas and electricity trade. Norway will obviously be a big player, with Russian gas and the electricity trade closely intertwined. To an extent it will follow the progressive liberalization of electricity and gas networks. Ultimately liberalization and emissions trading could become part of the *acquis communautaire* for the projected eastward enlargement of the European Union.

Over time all these processes will in some way become linked to one another, and one should perhaps be thinking about whether this has any implications for the negotiations on the second phase of the Energy Charter Treaty, which is the phase designed to deal with the terms of new investment, particularly in central and east European energy.

Conclusions

Finally, what is the role of Kyoto in all this? The Kyoto agreement cannot, need not and should not define internal national trading implementation systems. That does not matter, providing there is some international compatibility in the systems established, and an integrity verified for the national systems. In my view, the Kyoto agreement needs to define commitments that allow for the possibility of international and perhaps intertemporal trading and banking within fairly closely defined limits.

The Kyoto protocol needs to define the role of joint implementation and crediting within Annex I. As a final leg, it needs to establish a body which can do the jobs that are outlined here, namely certify that national systems will lead to compliance, and the compatibility of national permit systems and their integrity, as well as establish relationships to Joint Implementation where countries choose to use those instruments.

Just to remind us where we started, there are political attractions to emissions trading related to the negotiating structure of the emissions cap, related to the industrial politics of creating an asset and buying off certain quarters, and its institutional basis as an environmental measure rather than something which falls with the tax or other ministries.

Inevitably one will start with carbon dioxide. There are possibilities for expanding the regime over time, and the protocol negotiations need to think fairly carefully about whether and how they want to create a 'two lists' or other system that could move the structure towards gradually encompassing a more comprehensive approach.

Emissions trading along the US sulphur model would cover only some sectors – and for some countries, only minor sectors in terms of total CO_2 emissions. Other sectors could be addressed through sectoral targets, agreement on specific policy measures, and variants of trading systems applied higher up the energy stream, for example to refineries and fuel trading companies.

The details of trading systems are likely to vary considerably between different countries. The situation in Europe is of particular interest, given the failure of previous efforts and the current predicament, but it would require some strong and imaginative political leadership to overcome the current aversion to emissions trading. Central Europe is likely to be a key area for the development of Joint Implementation and its possible evolution towards permit trading.

Thus we will see varied national systems in which the key objective, if we are to attain the benefits identified, will be to make sure that international trading between those systems can evolve. Likewise we will see the national systems evolving. In that context the Kyoto protocol will not try to establish a global or Annex I system, but it will create incentives through the binding commitments, the enabling mechanisms that I have described, and the verification institutions, to ensure that emission trading happens in a credible way.

Chapter 15

Design and implementation of pilot systems for greenhouse gas emissions trading: lessons from UNCTAD

*Frank T. Joshua**

Introduction

In 1991 the UNCTAD secretariat initiated a research programme to examine the design and implementation aspects of a global emissions trading system for greenhouse gases, in anticipation of the adoption in 1992 of the Framework Convention on Climate Change. Since then, the UNCTAD secretariat has published five reports on this subject. In November 1996 the secretariat released a new study (and an illustrated synthesis) on the legal and institutional aspects of a pilot international GHG emissions trading system.

The new report on legal and institutional issues represents a major advance in our understanding of issues essential to the early establishment of a pilot GHG emissions trading programme. Prospects for the adoption of appropriate enabling provisions on cross-border emissions trading in the context of the FCCC have significantly improved following the adoption in July 1996 of the Geneva Declaration by the Second Conference of Parties to the FCCC (COP-2).

Some lessons from UNCTAD's R&D project

UNCTAD's work on GHG emissions trading is underpinned by three basic assumptions:

(1) Climate change is a global problem that will affect developed and developing countries alike. All countries will therefore need to contribute to the solution, though the developed countries must bear the

---

*   The views expressed in this paper are the author's and do not necessarily reflect those of the UNCTAD secretariat.

primary responsibility for these efforts (not only are they mainly responsible for the present problem of global warming, they are also best placed to bear the burden associated with finding solutions to it).

(2) Climate change raises fundamental issues of industrialization and development. These processes remain incomplete in developing countries. Global cooperative solutions to climate change must recognize the rights of developing countries to pursue their industrialization and sustainable development and, by implication, must accept increased emissions of greenhouse gases by those countries, especially carbon dioxide, for the foreseeable future.

(3) A global greenhouse gas emissions trading system would leave each country free to choose its own domestic policy mix for controlling GHG emissions at the national level (emission taxes, charges, regulations, etc.).

Early results from this research project helped to confirm some widely held conclusions, such as those relating to flexibility and cost-effectiveness, and pointed to other interesting observations. These include:

• The history of emissions trading in the United States provides ample evidence that emissions trading systems can achieve environmental goals at minimum cost. The US programme for the reduction of lead in gasoline (launched in the mid-1970s), the sulphur dioxide trading programme, and the Los Angeles RECLAIM (Regional Clean Air Incentives Market) programme are among the best-known examples. Some estimates put cost savings achieved by allowing trading under the US Clean Air Act at over $10 billion, not including recurring savings in operating costs.[1] Compliance under the SO_2 trading programme is well ahead of schedule at costs substantially lower than expected. As much as $1 billion may be saved in the implementation of SO_2 provisions with fully tradeable emission allowances. And as much as $58

---

[1] Tom Tietenberg and David Victor, 'Administrative structures and procedures for implementing a tradeable entitlement approach to controlling global warming', in *Possible Rules, Regulations and Administrative Arrangements for a Global Market in CO₂ Emission Entitlements* (in UNCTAD/GID/8, Part I, Geneva, 1994).

million a year (or some 40 per cent of the cost of using strict standards without trading) may be saved under the Southern California programme that allows trading in SO_x and NO_x.

- By enabling cheaper and more flexible compliance, emissions trading encourages more environmental benefits at an earlier stage. In effect, trading increases the political acceptability of larger emissions reductions.

- Emissions trading systems provide dynamic incentives for technological innovation and the transfer of new emissions reduction technology to poorer countries. Thus, properly designed, an emissions trading system could produce a 'triple dividend' in the form of economic, environmental, and technological benefits.

- While issues of equity and fairness should guide the choice of emission allowance allocation techniques, acceptable cooperative outcomes can be assured. Acceptable allocation techniques range from the traditional 'grandfathering' method (allocation based on current or historical emission levels), to efficiency principles (based on emissions per unit of GDP), egalitarian principles (per capita), economic differentiation (level of development), and weighted combinations of the above.

- If the emissions trading market is competitive and transaction costs are low, the initial allocation method will have no material bearing on the efficiency of the system to achieve a given environmental target at minimum cost. Therefore, the initial distribution of allowances can be used to address distributional and equity concerns without affecting the cost-effectiveness of the system. It is therefore possible to choose an allocation method that would reconcile the need to accelerate industrialization in poorer countries (for whom the exploitation of cost-effective abatement opportunities would not normally be a high priority) with the need to reduce emissions globally.

- The issue is not who buys or sells emission allowances; emissions trading produces win-win situations for both buyers and sellers. In addition to sharing in global abatement benefits, buyers will gain from the opportunities to meet their compliance obligations at lower cost than if they were to reduce emissions within their own borders. Sellers would benefit from their ability to undertake deeper emissions reductions than required and profit from the sale or retention of the excess allowances.

- A critical advantage of emissions trading over other emissions abatement policy instruments is its capacity to minimize transaction costs. Effectiveness in minimizing transaction costs requires reliance on market institutions (for efficient trading services), and on national authorities (for monitoring, certification and enforcement of the trading system).
- Adequate regulation is the foundation of an efficient emissions trading system (over-regulation could increase transaction costs).
- Ultimately, the credibility of any emissions trading system rests on the integrity of its monitoring, certification and enforcement institutions.

How to implement a GHG allowance trading system

This section draws on the results of UNCTAD's most recent research on legal and institutional aspects of a pilot international GHG emissions trading system. The principal issues to consider in the design and implementation of an international GHG emissions trading system are:

(a) basic principles of design and implementation;
(b) elements of a phased development;
(c) institutional structures; and
(d) financing of initial and recurrent operations.

(a) Basic principles of design and implementation.

- An international emissions trading system must be in the context of or pursuant to the FCCC.
- Eligibility should be open to countries adhering to legally binding emission targets and timetables within the context of the FCCC (this would include but would not be limited to Annex I parties).
- Participation in the trading system must be on a voluntary basis.
- Crediting among participants in the emissions trading system must be officially recognized.
- In order to minimize risks and the possible negative consequences of innovation, an international GHG emissions trading system should be constructed and implemented in a gradual, step-by-step approach.

(b) Some elements and benefits of a phased developmental path

- Building a global GHG emissions trading system in stages would facilitate the process of 'learning by doing'. This would ensure that the construction of the system is underpinned by empirical research and development activities helping to minimize risks and the potentially negative consequences of innovation.

- In a phased development of the emissions trading system, conventional policy instruments would continue to be utilized alongside innovative trading mechanisms, thereby increasing governments' flexibility in meeting their abatement obligations.

- On the basis of allowances trading, the development programme would allow for the gradual expansion of the trading system, starting with a limited number of countries, gases, industrial sectors, emission sources and sinks.

- Phasing the inclusion of countries, gases and sequestration projects into the construction of an international GHG emissions trading system is not only an issue of technical complexity but, more importantly, one of ensuring the integrity of the emissions trading market. Lack of consistency in the application of common protocols (trading, accounting, reporting, monitoring, certification, enforcement, etc.) would lead to commodity differentiation and market segmentation. This would severely undermine the emerging emissions market. Concern for predictability, uniformity and homogeneity of the commodity should therefore take precedence over the search for comprehensiveness of membership, gases, sources and sinks.

- A phased development of the system would also enhance transparency, raise the level of comfort and help to build confidence in the system.

The above suggests that a pilot GHG emissions trading system should be initiated along the following lines:

Phase One – Duration: 5 years (2000–2004)

- An emissions trading 'Group Agreement' should be negotiated among eligible countries on a voluntary basis, pursuant to the FCCC and its protocols.

- Eligibility to participate in the 'Group Agreement' should be open to countries adhering to legally binding emission reduction targets and timetables. This would naturally include but should not be limited to Annex I parties.
- The number of greenhouse gases and industrial sectors entering the trading system in phase one should be limited, and be based on the capability to accurately monitor or infer emissions and reductions. This could include energy-sector CO_2 emissions from major fixed-emission sources.
- A schedule for the expansion of the system to include other key GHGs (e.g., energy-sector methane) should be agreed in order to remove incentives for strategic gaming, such as shifting from coal-fired plants to natural gas with the potential for significant methane leakage. A schedule for the inclusion of sequestration projects would also help to encourage investments into that sector.
- Trading should be in emission allowances only, as trading of AIJ credits from emissions reduced or sequestered was disallowed by the Parties to the FCCC in their 1995 Berlin decision, and indications are that this provision could well continue beyond 1999.
- The issuance of emission allowances to participants in the trading system should be calculated on a cumulative basis, but allowances should be issued on an annual basis. This would allow for annual accounting and control of the system while providing longer-term flexibility, predictability and assurance for governments and industry alike.
- National institutions should allocate allowances internally, as well as providing monitoring, certification and enforcement services for the system (in order to maximize efficiency and reduce transaction costs).
- Private-sector institutions and neutral bodies should administer the market, conduct trades, transfer titles, pay and settle accounts, etc.
- Overall policy guidance, market regulation and supervision should be the responsibility of the participants in the Group Agreement.
- Interim international mechanisms should be established for (a) policy review, enforcement and regulatory functions, and (b) monitoring and certification functions.

Phase Two – Duration: 10–15 years (2005–2019)

- Plan for the inclusion of new participants (non-Annex I parties should be allowed to negotiate emission caps and timetables).
- Add new gases (such as energy-sector methane) and industrial sectors (e.g., large mobile sources such as shipping and air transport), sequestration projects (assuming the issue of crediting can be satisfactorily resolved, at least partially among participants in a voluntary trading programme). AIJ experience would be valuable in forging agreement on common protocols.
- Continue to rely on national institutions for domestic allocation of allowances, monitoring, certification and enforcement.
- Strengthen the role of the private sector by encouraging the development of information services and market promotion activities.
- Strengthen the roles of the interim international mechanisms charged with conducting high-level policy review, regulatory and enforcement functions, as well as with monitoring and certification.

The continued expansion and development of the emissions trading system will depend not only on the experience gained in previous phases, but also on technical progress in measuring and understanding the characteristics of various greenhouse gases and their interrelationships.

(c) Institutional requirements at the international level

If, as proposed above, a pilot international GHG emissions trading system were to rely on national institutions for such essential services as the allocation of allowances among domestic sources, monitoring, certification and enforcement of the system, then at the *international level* appropriate institutional structures would be required to:

- review and determine overall emissions allowance policy
- regulate the emissions trading market
- conduct and administer trading
- monitor emissions
- certify overall compliance with obligations
- enforce the system.

Effective execution of these functions would require a combination of three institutional pillars:

(i)   Policy review, decision-making, market regulation and enforcement should be implemented via a new intergovernmental body – the International Emissions Trading Organization (IETO).

(ii) The functions of monitoring and certification should be implemented via an independent technical body (to be established by participants pursuant to the FCCC), which may be referred to as the Monitor Institute.

(iii) Emissions trading, market administration, internal regulation, the delivery of trade-related services, etc. should be implemented via the private sector.

(i)   Main functions of IETO services related to trading[2]

- Issuance of allowances
- Issuance of credits
- Organizing auctions
- Recording trades
- Tracking allowances/credits
- Book-keeping/clearinghouse
- Licensing commodity exchanges
- Data services/trade information
- Software development.

  Other services:
- Market regulation
- Policy review and decision-making
- Enforcement.

---

[2]   IETO would have the capacity to delegate authority for some of its operational activities (e.g. the licensing of commodity exchanges) to a private sector or neutral market administrator.

(ii) Main functions of the Monitor Institute

- Monitoring and surveillance of emissions
- Certification of tradeable allowances/credits
- Reporting
- Preparation of common protocols/standards, etc.
- Dispute resolution.

(iii) Main functions of the private sector

- Trading of allowances/credits
- Conducting auctions on behalf of IETO
- Transfer of titles
- Payment and settlement of accounts
- Transfer of funds
- Provision of price and other market-related information
- Market-related services and software development, etc.
- Market administration, internal market regulation and supervision.

(d) Financing the initial and recurrent operations of an international GHG emissions trading system

An important feature of any emissions trading system is its ability to generate financial resources. Several of IETO's principal functions could result in activities that would be revenue-generating, such as the issuing of allowances and credits, licensing of commodity exchanges, trade information services, and software development. This could allow IETO to operate on a self-financed basis (after allowing for the underwriting of the initial development phase by bilateral sources, international institutions, NGOs, and private donors). IETO's recurrent operating costs should therefore have no budgetary implications for participating governments. However, the cost of operating the Monitor Institute would need to be absorbed by its member states. In principle, some resources from IETO's activities could also be earmarked for this purpose.

The UNCTAD secretariat has been encouraging public/private partnerships to implement a pilot greenhouse gas trading programme and is working with the Earth Council and Centre Financial Products to develop a pilot emissions market through the establishment of the Global Environmental Trading System (GETS).

Conclusion

Progress in developing AIJ projects and in introducing emissions taxes has fallen far short of expectations, while recent successes with emissions trading programmes in the United States (SO_2 and RECLAIM) are encouraging new interest in the development of cross-border GHG emissions trading. It is evident that such trading could provide significant environmental and economic benefits.

In the long run, because of comparatively lower transaction costs, emissions trading systems may progressively crowd out AIJ activities. However, scheduling the inclusion of emissions reduction and sequestration credits into a trading system should provide further incentives for early investments in those projects.

AIJ and emissions trading should be regarded as complementary mechanisms working towards the same environmental goals. Their co-existence is necessary and desirable if early and substantial abatement results are to be achieved.

References

Robin Clarke, *A Pilot Greenhouse Gas Trading System: The Legal Issues* (in UNCTAD/GDS/GFSB/Misc.2, Geneva, 1996).

Robin Clarke, *Controlling Carbon Dioxide Emissions: The Tradeable Permit System* (in UNCTAD/GID/11, Geneva, 1995).

Combating Global Warming: Study on a Global System of Tradeable Carbon Emission Entitlements (in UNCTAD/RDP/DFP/1, Geneva, 1992).

Richard L. Sandor, Joseph B. Cole and M. Eileen Kelly, 'Model rules and regulations for a global CO_2 emissions credit market', in *Possible Rules, Regulations and Administrative Arrangements for a Global Market in CO_2 Emission Entitlements* (in UNCTAD/GID/8, Part II, Geneva, 1994).

Richard B. Stewart, Jonathan B. Wiener, and Philippe Sands, *Legal Issues Presented by a Pilot International Greenhouse Gas Trading System* (in UNCTAD/GDS/GFSB/Misc.1, Geneva, 1996).

Chapter 16

Emission trading design options and environmental performance

Jan Corfee-Morlot[*]

Introduction

At the Second Conference of the Parties to the UN Framework Convention on Climate Change, Ministers issued a Declaration to endorse the findings of the IPCC's Second Assessment Report and instructed their representatives to accelerate negotiations on the text of a 'legally binding' protocol or other legal instrument under the Convention.[1]

Several Parties to the Convention have suggested that greenhouse gases (GHG) emission trading may be a viable option to achieve new commitments under the FCCC defining legally binding targets for GHG reduction.[2] The purpose of this chapter is to consider a few of the basic design issues that are likely to be key in considering the use of emissions trading as part of an internationally agreed mitigation response strategy. The chapter attempts to lay out some of the principal differences among alternative design options for international emission trading. It reviews two basic alternatives for emission trading schemes – cap and allocate versus baseline and credit – highlighting important differences among them, and moves on to consider the nature of the commodity that might be traded. It addresses the rather narrow question of environmental performance, considering a range of different basic designs for emission trading systems. Clearly there might be other criteria for policy-makers, but they are not considered here. Further, it could be argued that environmental performance should the overarching objective of any mechanism adopted under the Convention.

---

[*] The views presented in this paper are those of the author and do not represent the views of the OECD or of its member countries.
[1] UNFCCC Ministerial Declaration, 1996.
[2] AGBM/1996 Misc. 2, Add. 1 and 2; New Zealand, 1996.

This chapter adds a few more issues to an already complicated debate about the possible use of emission trading as an international policy tool to mitigate GHG. In doing so, the intention is to add a bit of practical thinking to what has been a theoretical discussion and is now becoming a highly politicized debate. Such practical considerations could help policy-makers think clearly about the alternative design options for an international emission trading system and the ability of these alternative systems to achieve various objectives under the FCCC.

Basic design options

A basic starting point for discussion on emission trading options is to clarify the range of alternatives for the basic structure of such a system. In the literature, especially that which draws on the US experience with emission trading, two different types of systems are identified: (1) cap and allocate and (2) baseline and credit.[3] These two types of systems may have quite different policy implications. In particular, the choice between these two alternatives may affect the ability of governments to rely on emissions trading as the principal policy instrument or simply as one of a broader package of instruments designed to work in parallel with one another, to achieve a set of collective national emissions reduction targets.

'Cap and allocate' compared to 'baseline and credit'

Cap and allocate and credit and baseline trading alternatives have some similarities but are in many ways different. When designed effectively, both provide improved flexibility compared to uniform emission reduction targets for the polluter to decide on how, when and where investments are made to achieve GHG reductions. Both alternatives also offer significant improvements in the cost-effectiveness of achieving a given level of GHG emission reduction.

The main difference between a cap and allocate system and a credit and baseline system in the GHG context is the certainty with which overall

---

[3] OECD, 1992a and Kete, 1996.

emission reductions among the different participants will be achieved. A cap and allocate system should ensure achievement of the cap and hence measurable, verifiable reductions in total emissions. Allowances are the unit of trade. Should the participant not require all of its emission allowances they may be 'traded' to another participant.

Credit and baseline systems, when discussed in the context of GHG mitigation, are often presented in the form of 'uncapped' emissions trading. This is not the only way to design a credit system; however, it may be the simplest form. In a credit system, the size of the credit is equivalent to the difference between the agreed baseline and actual emission levels. Either of these systems may include different types of participants and encompass some or all of the emissions from those participants.

If defined as a standard against which to measure progress, baselines in practical terms can be derived in a number of different ways. Some of the possibilities are:

- business as usual or 'without policy' baselines, project by project;
- business as usual or 'without policy' baselines for a firm or across an industry group, for a sector, or for a country;
- 'with policy' baselines for a firm, industry group, sector or country (this might be a performance standard or, alternatively, a total emissions level lower than what is expected to occur without policy intervention).

Only 'with policy' baselines are sure to lead to a reduction in GHG beyond what would have happened without implementation of the policy. Figure 16.1 is a schematic diagram showing two different types of baselines ('without policy' versus 'with policy') and their respective credits. If one takes the target, represented by the baseline in line B, as the desired level of emissions, under Case A the actual emission level, after creation of credits, achieves this level of environmental performance. Alternatively, Case B has a 'with policy' baseline which begins at this point. Emission credits are earned if the participants exceed the desired level of performance. Action to generate credits is voluntary in both instances.

Business as usual baselines also present methodological concerns on

Figure 16.1: Credit and baseline: alternative definitions of 'baseline'

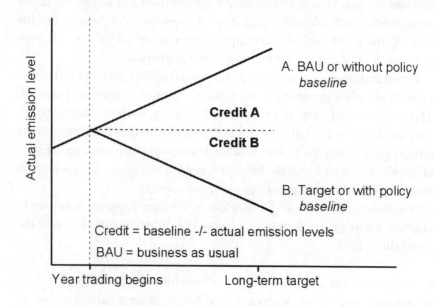

Sources: Arquit-Niederberger, 1996 and Storey, 1996.

how to assess the validity of the baseline and hence how to document whether real environmental gains have been made relative to what would have happened without the transaction. Conceptually, the methodological issues are similar to those encountered in the development of consistent reporting and monitoring procedures for 'activities implemented jointly' or even for voluntary agreements.

Emission credit trading in the United States operates under broad, well-established regulatory frameworks.[4] 'Credits' in these systems have often been measured against agreed performance standards for products or emission sources. Lead in gasoline trading in the United States offers a good example of how such a system can work.[5] The system, which was designed to provide flexibility and reduce costs of achieving regulations on the lead

---

[4]   Unterberger, 1995.
[5]   See Figure 16.2 and Nussbaum, 1992.

Figure 16.2: Predicted and actual lead concentration under banking

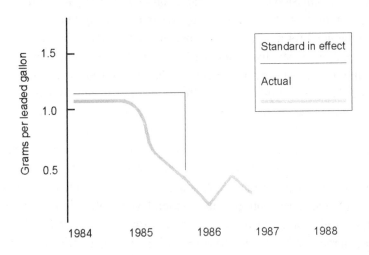

Source: Nussbaum in OECD, 1992.

content in gasoline, centred on agreement on a schedule for an increasingly stringent standard, phased in over three years. The participants were oil refiners or producers of gasoline. Figure 16.2 shows how over-achievement in the early period, when emission standards were relatively lax, was used to create valuable emission reduction credits. Part of the value of the credits derived from the ability to use them up to three years later. The system rewarded over-achievement by providing a financial incentive for the creation and marketing of such credits. At the same time, the system provided flexibility to emitters to achieve the agreed standards through purchase of credits should those credits prove to be less costly than direct investment in cleaner production processes or technologies.

If this system works, 'everyone' is happy: the regulators are happy because standards have been established and will be met; the environmentalists are happy because environmental performance has improved; the emitter is happy because the cost of meeting the environmental standard

has been reduced, and in some instances companies may have made profits through the marketing of this new commodity of 'emission credits' or
excess avoided emissions. But the satisfaction of all participants is dependent on a common starting point for the design of the system: everyone
agrees on the need for a performance standard with a policy baseline, and
to work together to establish a standard that will be environmentally significant; this is followed by agreement on the rules for emissions trading
to enhance the 'flexibility' and cost-effectiveness of means to achieve it,
while not compromising on the overall agreed environmental target.

Does this or could this experience have parallels within the AGBM
Annex I Group? The Commission of the European Communities is a case
in point, as an international governmental organization that has tried to set
agreed international performance standards for major appliances (white
goods) under the SAVE programme. Several years after initiating this
Programme, only three out of fifteen of the originally targeted end-uses
have been agreed on.[6] By comparison with US appliance efficiency standards, the EC standards are quite weak.[7] If this is the outcome of the negotiation among relatively homogeneous EC member states, what would be
the outcome of a similar undertaking among the relatively diverse set of
Annex I countries?

Capped versus uncapped and mixed systems

The foregoing discussion indicates that a basic difference between credit and
allowance systems is that credit systems may be 'uncapped' trading of emission reduction credits while allowance systems are by definition 'capped'.
However, one can also imagine the design of a fully capped yet credit-based
emission trading system. Dan Dudek has suggested that caps combined with
credit trading might be possible.[8] He suggests that emission budgets by
country would be the cap for the individual national system; in his view
credits would be created only after reductions had been achieved, verified
and documented. These credits could then be traded internationally.

---

[6]   CEC, 1996.
[7]   Mullins, 1996.
[8]   Dudek, 1996.

However, he also advances the suggestion that uncapped trading, based on performance measured against project baselines in countries without agreed 'budgets', could occur in parallel with such a capped system. This would mean that credits from both capped and uncapped systems would be mixed in an international market and traded internationally. A simple question concerning environmental performance is relevant: how would one design such a system to ensure GHG reduction results? That is, if all the investment in mitigation were to migrate to nations that had not adopted national baselines, how would one verify overall progress in GHG reduction?

It is also possible to imagine using a GHG or CO_2 intensity or performance standard, or series of standards, to cover the range of different sources. However, unlike lead emissions, CO_2 emissions alone have a variety of different kinds of sources. Even if one were able to agree on performance standards for some of the major emitting sectors or kinds of sources (say industrial or power generation facilities), what is to prevent overall emission growth in other sectors from outweighing the benefits achieved in these sectors? One solution might be to use different kinds of tools used in tandem to achieve simultaneous reductions in different kinds of sources. Tietenburg and Swart have commented on this, suggesting that measures other than emissions trading might be best adapted to sectors characterized by a great number of small sources and diversity of different kinds of actors.[9] Mixing other policy instruments with emission trading might be essential to achieve agreed national targets, especially if 'uncapped' trading systems are favoured by policy-makers.

Defining the commodity or unit of trade

Some have suggested mixing different kinds of sources and sinks of CO_2 to take advantage of the significant differences in marginal costs of mitigation for each. Similarly, an argument can be made that comprehensive systems, covering all different GHG (CO_2 sinks and sources, CH_4 and N_2O and precursors) provide even greater opportunities for cost savings to achieve a given target.

---

[9]   Swart, 1992.

Table 16.1: Estimated ranges of uncertainty for GHG inventories (%)

Source/sink	Swart	UK	USA
CO_2 energy	±10	±2–5	<10
CO_2 deforestation	±35	±20	±25
CO_2 other & forestry	±25	±20	±25
CH_4 energy	±30	±20	±10–25
CH_4 landfills	±50	±20	±10
CH_4 animal husbandry	±20	±20	±10
CH_4 from rice	±40–60	±20	±10–25
N_2O energy	±50	±25	±25
N_2O agriculture/forestry	±50–100	±25	±25

Sources: UK, 1994; USEPA, 1994; and Swart in OECD, 1992a.

One of the key issues in the creation of a credible trading system is that the commodity to be traded must be easily measured, hence enabling monitoring and verification (such as independent audits).[10]

Table 16.1 shows the wide range of uncertainty associated with different sources and sinks of GHG as estimated by Swart and government analysts in the UK and the United States. Only the inventories of CO_2 from energy are estimated to have less than 10 per cent uncertainty. Estimates of CO_2 from deforestation are often associated with more than double the uncertainty range of CO_2 from energy. The uncertainty of CH_4 inventories varies by source, but also tends to be higher than that for CO_2 from energy. Similarly, N_2O may be more than double that of energy and with one analyst suggesting a range as high as 100 per cent. Uncertainty estimates of national inventories among different sources and sinks of GHG might be an indicator of the ability to measure emissions from these activities. If so, then it would be difficult to argue, at least currently, for a fully comprehensive system.

---

[10] Unterberger, 1995 and Nussbaum, 1992.

Table 16.2: Basic design options and their outcomes

Likely outcome/principal objective	Other results	Basic design options	Examples
Achieve agreed emission reduction targets with high certainty and cost-effectively	Establish incentives/market for clean technology and exceeding agreed emission reductions	'true' baseline cannot be measured	Cap/allocate – country level 'With policy' country baselines / credit
	Improved cost-effectiveness of emission reductions		
Cost-effective improvement in the GHG intensity of participants	Progress towards overall national GHG targets; impossible to achieve national targets uniquely with this instrument		Sector, industry caps/allocate 'With policy' baselines for the sector or industry segment of a sector
Establish markets or incentives for clean technology	Cost-effective GHG emission reduction should result; difficult to know with certainty emission reduction results since		'Without policy' baseline/credits for countries, sectors, industries, firms, projects (uncapped system) and crediting.

Source: Author.

Is environment the bottom line?

Table 16.2 distinguishes some of the basic design options for emission trading according to likely outcomes or system objectives. The systems are ranked according to the expected environmental performance, with the first row representing the highest performance and the bottom row the lowest performance.

Each of the basic system designs should have different strengths and weaknesses. This chapter has not attempted to identify fully the different strengths or weaknesses of the alternatives, nor to debate the value of different possible outcomes or policy objectives. Other objectives and outcomes may be just as valuable as the rather narrow question of environmental performance addressed here. For example, initially it may be necessary for any agreed emissions trading system to demonstrate that it can play a role in GHG reduction, and the objective may be to gain experience with this new tool internationally. If this is the case, then it will probably play a smaller role in national efforts to achieve agreed targets than if the objective is to use the system as a tool to achieve national targets in the most cost-effective manner across the set of international players. Clearly identifying these objectives at the outset will help to assure the success of the programme.

Any new agreement to reduce GHG under the FCCC should produce verifiable environmental results. Strengthening obligations under the FCCC means that collectively Parties agree to reduce GHG emission levels below what would have occurred without such an agreement. In the design of an emission trading system policy-makers need to ask: whether environmental performance is the objective and likely outcome for this system. There may be strong reasons why it is not the bottom line in a first-generation international emission trading system. If this is the case, however, the system should be developed in a more limited context. This may seem like an obvious point, but unfortunately some advocates of the credit form of emissions trading schemes forget, in their search for cost-effectiveness, to address the question of whether such schemes can be proved to have made a difference to the environment.

References

Arquit-Niederberger, Anne (1996). 'Activities Implemented Jointly: 'Review of Issues for the Pilot Phase'.

CEC (1996). National Communication from the Commission under the UN Framework Convention on Climate Change, Brussels, 11 June 1996, COM(96) 217 Final.

Dudek, D.J. (1996). 'Emissions Budgets: Creating Rewards, Lowering Costs and Ensuring Results', paper presented to the Climate Change Analysis Workshop, Springfield, Virginia, June 6–7 1996.

Dwyer, John P. (1992). 'California's Tradeable Emissions Policy and its Application to the Control of Greenhouse Gases', in OECD (1992a).

Kete, Nancy (1996). 'Twenty-five years of experience with emissions trading programs', presentation to the OECD Forum on Climate Change, Paris, 30 September–1 October 1996.

Kete, Nancy (1992). 'The US Acid Rain Control Allowance Trading System', in OECD (1992a).

Mullins, F. (1996). 'Energy efficiency standards on tradeable goods', Policies & measures for common actions working paper, OECD, Paris.

New Zealand (1996). *Climate Change and CO$_2$ Policy: A Durable Response* (discussion document of the Working Group on CO$_2$ policy), New Zealand Ministry of Environment.

Nussbaum, Barry D. (1992). 'Phasing Down Lead in Gasoline in the US: Mandates, Incentives, Trading and Banking', in OECD (1992a).

OECD (1995). *Global Warming: Economic Dimensions and Policy Responses*, OECD Publications, Paris.

OECD (1993). *International Economic Instruments and Climate Change*, OECD Publications, Paris.

OECD (1992a). *Climate Change: Designing A Tradeable Permit System*, OECD Publications, Paris.

OECD (1992b). *The Economic Costs of Reducing CO$_2$ Emissions*, OECD Economic Studies, OECD Publications, Paris.

Palmisano, John (1996). 'Air Permit Trading Paradigms for Greenhouse Gas: Why Allowances Won't Work and Credits Will', personal communication.

Storey, Mark (1996). 'Policies and Measures for Common Action, Demand Side Efficiency: Voluntary Agreements with Industry'.

Swart, Rob (1992). 'Greenhouse Gas Emissions Trading: Defining the Commodity', in OECD (1992a).

UK (1994). *Climate Change: The UK Programme, United Kingdom's Report under the Framework Convention on Climate Change*.

UNCTAD (1995). *Controlling Carbon Dioxide Emissions – The Tradeable Permit System*, United Nations Conference on Trade and Development, United Nations, Geneva (UNCTAD/GID/11.40).

UNCTAD (1994). *Combating Global Warming: Possible Rules and Regulations and Administrative Arrangements for a Global Market in CO_2 Emission Entitlements*, United Nations, Geneva (UNCTAD/GID/8).

UNFCCC (1996). The Ministerial Declaration made by the FCCC on behalf of the Ministers and other heads of delegations present at the second session of the Conference of the Parties to the United Nations Framework Convention of Climate Change Geneva, Switzerland, 17 July.

Unterberger, Glenn (1995). 'Turning Emissions Reductions into Fiscal Assets - Can Environmental Regulations Generate Income as Well as Costs?', *Journal of Environmental Regulation*, Spring.

USA (1996). Ministerial Statement made by the Hon. Timothy E. Wirth, Under Secretary for Global Affairs to the Second Conference of the Parties, Framework Convention on Climate Change, Geneva, Switzerland, 17 July.

USEPA (1994). Inventory of US Greenhouse Gas Emissions and Sinks: 1990–1993. USEPA, Office of Policy Planning and Evaluation (2122).